SpringerBriefs in Plant Science

For further volumes:
http://www.springer.com/series/10080

Natalie G. Mueller

Mound Centers
and Seed Security

A Comparative Analysis of Botanical
Assemblages from Middle Woodland
Sites in the Lower Illinois Valley

 Springer

Natalie G. Mueller
Department of Anthropology
Washington University in Saint Louis
St. Louis, MO, USA

ISSN 2192-1229   ISSN 2192-1210 (electronic)
ISBN 978-1-4614-5920-0   ISBN 978-1-4614-5921-7 (eBook)
DOI 10.1007/978-1-4614-5921-7
Springer New York Heidelberg Dordrecht London

Library of Congress Control Number: 2012950778

Printed on acid-free paper

Springer is part of Springer Science+Business Media (www.springer.com)

Contents

Mound Centers and Seed Security: A Comparative Analysis of Botanical Assemblages from Middle Woodland Sites in the Lower Illinois Valley

Contents

The earthen mounds built by the Middle Woodland inhabitants of the Eastern Woodlands have been the focus of archaeological research for more than a century. Within these mounds, excavations have revealed naturalistic art worked on exotic materials from points as distant Wyoming, Ontario, and the Gulf Coast (Carr 2006b). At the turn of the twentieth century, the makers of this 2,000-year-old art and architecture were named the Hopewell culture and envisioned as a cohesive and highly sophisticated society inhabiting southern Ohio (Moorehead 1892). In the century since, Hopewell has been transformed into a descriptor of a complex network of exchange and interaction spanning the river valleys of the Eastern Woodlands.

Concurrent with this interpretive shift, paleoethnobotanical research has shown that Middle Woodland societies produced crops of native seeds (referred to as the Eastern Agricultural Complex) before the introduction of maize to eastern North America. This study examines the botanical remains recovered from the Mound House site in the Lower Illinois River Valley (LIV), one point of articulation in the network of Hopewellian interaction. I compare the patterning of plant remains at

Mound House to that reported for other mound and habitation sites in the LIV. I argue that differences between these assemblages reveal a relationship between Hopewellian ritual and exchange and the emergence of eastern North America's first agricultural system during the same period.

Today, the Mound House site is unobtrusive; a low mound nestled into the endless cornfields of the Illinois River Valley, dwarfed by the network of levees that surrounds and protects it. The levees define modern life in the valley, which is dominated by mechanized, industrial farming on a massive scale. This economy takes its inputs from factories and laboratories and exports its products to markets spanning the globe. Mound House is a remnant of an older, obscured landscape and a node in smaller, slower network. Archaeological research on the Middle Woodland (c.100 BCE–500 CE) settlement of the LIV, as one expression of the Hopewell phenomenon, has always been focused on articulating this landscape and the network of human activity that stitched it together. In 1964, Stuart Struever identified the primary postulate of Hopwellian studies that "...some form (or forms) of communication, intercourse, or articulation existed prehistorically to enable far-distant groups to share an assemblage of imported raw materials, artifact styles, and precepts governing the internment of certain dead" (1964: 106). To this list of shared materials and ideas, I would add *seeds* and *agricultural knowledge*.

The Middle Woodland period in the midwest and midsouth is marked by the emergence of agricultural systems and by the appearance of networks of elaborate earthworks and trade goods (Carr 2006b; Charles 2006; Smith 1992a). How were these two phenomena connected? This chapter will ask what role mound centers played in the dissemination and intensification of agricultural systems during the Middle Woodland period, using Mound House as a case study. As monumental constructions, it is clear that mound centers were community projects. As the locations of iterative ritual construction over a period of hundreds of years, mound complexes were places that people returned to. They provided a formal time and place for exchange and so guarded against the loss of cultural knowledge and materials in small-scale, relatively mobile, and fragmented societies. Riverside mound centers in particular were positioned to increase the chances of interaction and exchange between far flung populations.

Because of their integrative potential, the proliferation of mound centers during the Middle Woodland may have facilitated the acquisition of agricultural knowledge and material by individual households. Maygrass (*Phalaris caroliniana*) is one possible example. It is a wild grass that is not native to western Illinois and appears only very rarely in pre-Middle Woodland deposits in the valley (Asch and Asch 1985d: 169). At Middle Woodland sites, including Mound House, it is both ubiquitous and abundant. At some point, this seed must have been carried out of its southern habitat by people, along with the knowledge of where and when to plant it and how to harvest and prepare it. Sites like Mound House are a logical place to look for a locus of exchange.

But other plants that were cultivated during the Middle Woodland already had long histories of use in the LIV. Squash and gourd were cultivated by the Middle

Archaic, whereas seed crops such as marsh elder (*Iva annua*), goosefoot (*Chenopodium berlandieri*), giant ragweed (*Ambrosia trifida*), and sunflower (*Helianthus annuus*) were cultivated by the Late Archaic (Asch and Asch 1985d). For these crops, researchers seek to explain not the appearance of novel plants, but the striking difference in the concentration of seeds between the Late Archaic and the Middle Woodland. Here, mound centers were well placed to serve an integrative role on a much smaller scale. Ethnographic examples of local institutions for the exchange of seed stock provide insights into how small-scale farmers mitigate risks and improve crops (Badstue et al. 2006; McGuire 2008; Misiko 2010; Stromberg et al. 2010). I will use the botanical assemblage from Mound House to evaluate the hypothesis that mound centers served as institutional supports for the intensifying agricultural systems of Middle Woodland societies. Physically, mounds were focal points on the landscape; behaviorally, the iterative nature of construction and ritual at mound centers provided a reliable locus of communication and exchange for the inhabitants of the valley.

The Mound House site is centered around two extant mounds located on a sand bar approximately 400 m from the channel of the Illinois River. Historical surveys indicate that the mound complex originally consisted of three to five mounds, but much of the site has been obscured by the levee that now protects it (Fig. 1; King et al. 2010). The last published site report (Buikstra et al. 1998) focused on excavations in and around Mound 1 (the largest extant mound) and argued that (1) material culture recovered at Mound House does not support models which have divided Hopewell ritual and habitation sites into separate functional categories in the past and (2) both the floodplain location of the site and the use of special soils (inverted sod blocks) in mound construction can be used to argue that ritual at Mound House was focused on themes of world renewal with annual floods acting as a metaphor, an argument that will be reviewed in greater detail below (Buikstra et al. 1998; Charles et al. 2004; Van Nest 2006).

Starting in 2009, excavations shifted to an adjacent area on the sand bar where magnetometry surveys indicated substantial subterranean anomalies (Fig. 2). Excavators hoped to further investigate the possibility that the Mound House site crosscuts ritual and domestic functional categories and to elucidate the nature, duration, and intensity of habitation. Excavations revealed a large irregular pit (Feature 379) approximately 3 m long and 1 m deep, and other areas of concentrated midden, but have unearthed no definite structures as of 2011. Seventeen flotation samples (139.5 l of sediment) taken from Feature 379 have yielded a rich record of plant use at Mound House. In comparison with assemblages from other sites in the LIV, they can help clarify the function of floodplain mound centers within Middle Woodland anthropogenic landscapes.

This analysis adopts a comparative approach. I begin by situating the Lower Illinois River Valley within the Middle Woodland trends of the Eastern Woodlands, and proceed to "zoom in," placing Mound House within the valley, and finally examining the history and interpretation of the site itself. Next, I conduct a detailed comparison of the new botanical assemblage from Mound House with those known

Topography of the Mound House Site

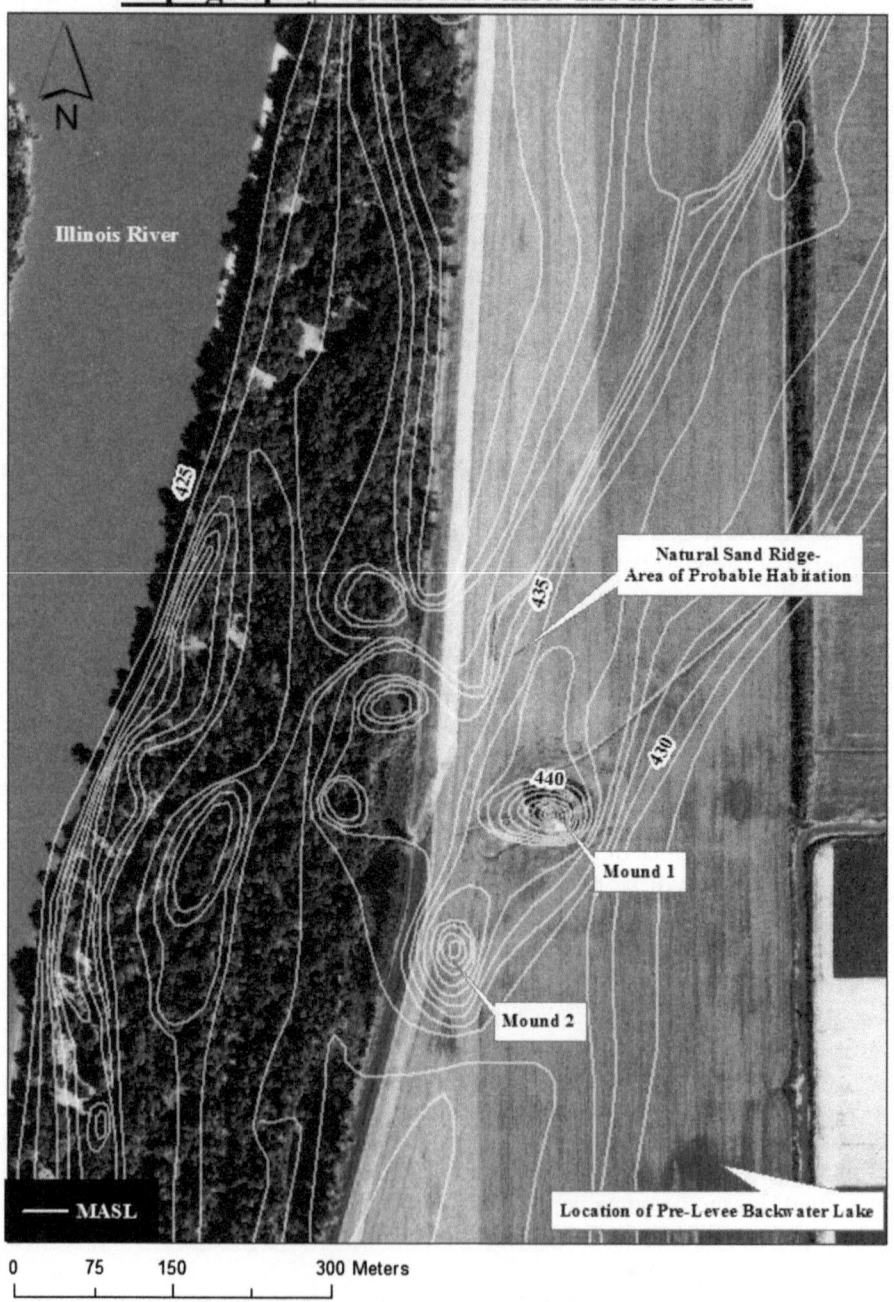

Fig. 1 This map shows the extent of the Mound House site. The topographic layer was created after a 1905 survey that predated the construction of the levee. A third mound is visible northwest of Mound 1; the two depressions south of it may have been borrow pits. The aerial photograph was taken in 1954 and shows the current location of the levee (between the field and the brush covered area on the western half of the site) and the road (cutting through the southeastern part of the site). Topographic data and photo courtesy of the Center for American Archaeology. Notice that the southeast corner of the site is only 3 m above the surface of the river in midsummer. The water level of the Illinois River fluctuates just over 6 m between low water and major flood stages at the Hardin gauge, the gauge closest to Mound House (National Weather Service Advanced Hydrological Prediction Service)

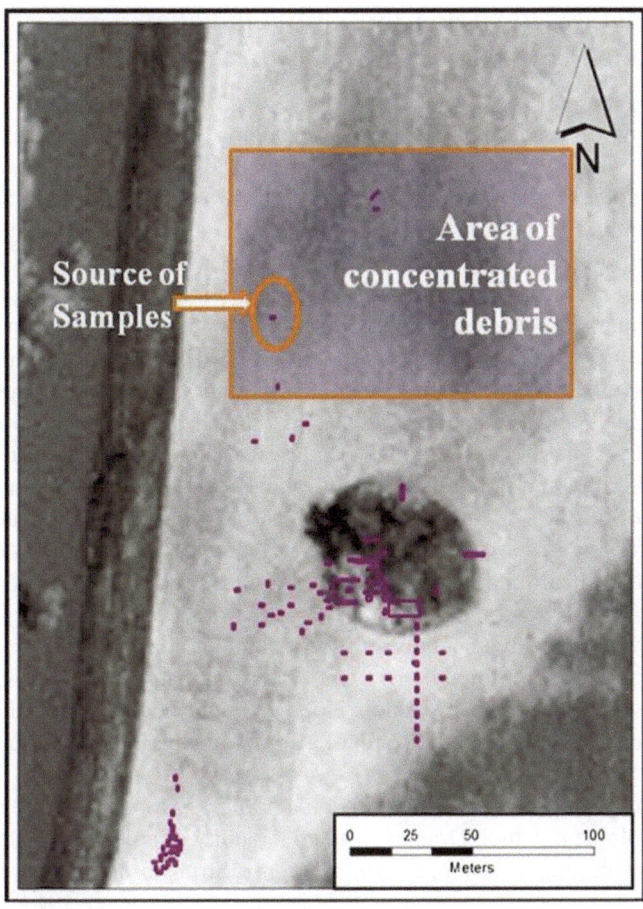

Fig. 2 Aerial photograph (1988) and map of excavations, centered on Mound 1. *Squares* show previous excavation units The cluster of units in the southwest explored the remnants of Mound 2. Feature 379, the source of the samples under analysis, is *highlighted*. Courtesy of the Center for American Archaeology

from five other sites in the valley, three habitation sites and two other mound complexes, to clarify the function of mound centers with regard to the emerging agricultural system of the Middle Woodland period. I conclude by reviewing some recent studies that describe the structure of seed exchange systems among modern subsistence scale farmers. These studies explore the variables that farmers take into consideration when deciding when, where, and with whom to exchange seeds and can provide insights into the possible role that mound visitations played in ancient seed networks.

What Is Hopewell?: The Lower Illinois Valley in a Regional Context

After a century of research on Hopewell archaeology, it is still necessary to redefine at the outset what Hopewell connotes, as do both of the most recent syntheses of Hopewell studies (Carr 2006a; Charles and Buikstra 2006). For the first half of the twentieth century, Hopewell was variously defined as a culture or a cultural complex. This use of Hopewell harks back to the first systematic studies of the Ohio Hopewell mounds around the turn of the century, which proceeded on the assumption that builders of Hopewell earthworks were separate from, and more advanced than, their village-dwelling neighbors (Shetrone 1920). A fairly uncomplicated connection was made between the elaborate art and architecture associated with mounded burials and the "culture" that apparently spanned large portions of the riverine midwest and south, although some researchers, especially Griffin (1952), stressed variability and noted the probability of local cultural developments. Other researchers of this era, such as Wray (1952) and McGregor (1958), interpreted the Illinois expression of Hopewell culture as derivative, resulting from an expansion out of a cultural heartland in southern Ohio. There was a tendency persisting into the 1960s to directly associate the suite of burial practices and trade goods called Hopewell with a cohesive social group or a single community on the move. This type of thinking is exemplified by Dragoo's assertion that "the similarities shared by Ohio Hopewell, Central Basin-Illinois Hopewell, and Point Penninsula [were not] merely the result of trade, but the result of a common physical and cultural heritage" (1964: 19).

Interestingly, the seeds of a new understanding of Hopewell were sown in the very same volume in which Dragoo asserted the unity of archaeological culture and real human community. In *Hopewellian Studies*, Struever and Caldwell redefined Hopewell as an "interaction sphere" (Caldwell 1964; Struever 1964). The key to their new terminology was a division between the interregional trade complex and the various local subsistence and settlement systems, each of which was envisioned as a distinct society. Struever (1964: 88) contended that the Interaction Sphere facilitated the movement of raw materials and ideologies that translated themselves into variable local stylistic expressions. For the Lower Illinois Valley itself, he points out that the dissemination of Hopewell style ceramics accompanies an elaboration of mortuary ritual but that burial practices were structured differently than those in the Ohio Hopewell area.

The Hopewell Interaction Sphere concept was extremely influential throughout the 1960s and 1970s and is especially well represented in the classic volume *Hopewell Archaeology: The Chillicothe Conference* (1979). The concept of an interaction sphere inspired many attempts to locate gateway communities or regional transaction centers, some more successful than others. These analyses were constrained by an incomplete understanding of settlement structure, subsistence, and utilitarian material culture, leaving connections to be made primarily on the basis of mortuary sites. Unfortunately, data from habitation areas was most sparse in the southern Ohio Hopewell heartland itself. Despite the lack of habitation site data, ecological models were often employed to explain local variation or change over time, but few researchers provide adequate evidence to prove or disprove such models. For example, Brose (1979) presented a "speculative model for the role of exchange." He hypothesizes that during the Late Archaic, societies began to specialize in local resources and form connections with other groups to hedge against local shortfalls. Over time the success of this system was self-reinforcing because larger populations were forced to specialize, and thus rely on exchange, more and more over time, culminating in the extensive exchange networks evidenced by Middle Woodland mounds.

Struever also posited an increase in population signaled by larger villages and deeper middens during the Middle Woodland. He attributed this increase to the emergence of what he referred to as "horticulture," based on new evidence derived from systematic flotation at excavations in the LIV. Struever saw early horticulture as intimately tied to the floodplain. As populations increased in response to improved food security and increasing sedentism, segments of society budded-off, carrying ideas and material down the river systems of the midwest (1964: 102–104). This model fit his study area, the LIV, very well, because this segment of valley seems to have been sparsely populated during the Early Woodland; migrants from the north and possibly from the south began to arrive in the lower valley at the beginning of the Middle Woodland period (Farnsworth 1990: 116; King et al. 2011).

As researchers put Struever's hypotheses to the test, they were able to demonstrate how much explanatory power can be derived from the integration of detailed settlement surveys and subsistence data on a local scale. Large-scale paleoethnobotanical analyses for the LIV published in the late 1970s and early 1980s cast doubt on the conventional wisdom that Middle Woodland settlement consisted of base camps on or near the floodplain and smaller procurement camps in the uplands along secondary streams (Asch et al. 1979). These studies refuted the "mudflat hypothesis" forwarded by Struever by pointing out that members of the Eastern Agricultural Complex (EAC) can be grown in the uplands and indeed may have alleviated the need to live close to the floodplain (Asch et al. 1979; Munson 1984). Importantly, these ecological analyses were based on data from upland sites that demonstrate the cultivation and processing of seed crops there, rather than on hypothetical scenarios. A new hypothesis could then be forwarded: that plant cultivation relieved packing pressure near highly concentrated floodplain resources by rendering less desirable (upland) locations more productive (Asch et al. 1979).

Struever was not the only one to see horticulture as a causal factor in the development of Hopewellian exchange. Richard Ford's (1979) elegant theoretical synthesis of intensifying plant cultivation is still useful today. He describes a dichotomy between centripetal and centrifugal subsistence systems. Centrifugal systems are those "based on the productivity of a natural ecosystem with some supplement from gardening" (1979: 237). Such a system is centrifugal in the sense that it requires people to range widely because important resources, such as nuts, are not available in the same place from year to year. To prevent local short falls, leaders need to schedule gatherings at specified times and places. Ford argued that Middle Woodland systems fell into this category and that the networks of trade and the elaborate public architecture associated with the Hopewell phenomenon were consequences.

According to Ford, agriculture tends to increase community self-sufficiency, creating a centripetal social system defined by the growth of local institutions for aggregation and exchange. Ford (1979: 238) saw centripetal systems developing in the Late Woodland and Mississippian, citing a lack of evidence for fallow field plants in Middle Woodland refuse. But more recent reviews of the literature from the American Bottom show a remarkable increase in hazelnut shell in Middle Woodland middens (Simon and Parker 2006); the same is also true for West Central Illinois (Asch and Asch 1985a: 351–353). Hazels are fire-tolerant colonizers of forest openings and are likely to increase on a landscape where clearance by fire and fallowing are practiced. Over thirty years of additional paleoethnobotanical evidence also indicates that crop production had begun by the Middle Woodland period, contrary to Fords original model.

Nevertheless, Ford's model is useful but seldom used. Ford emphasizes the importance of certain places as venues for *formal visitations*; these resulted in institutions for exchange. Instead of exemplifying a centrifugal system, as Ford believed, Middle Woodland societies might be conceptualized as in transition from a centrifugal to a centripetal form. If this is the case, we would expect to find locations dedicated to formal visitation on both a regional and local scale and to see local institutions for communication and exchange becoming increasingly elaborate over time. The history of mounds and earthworks as locales for regional visitation and trade extends far back into the Late Archaic. I argue that it is the *local* function of the mound centers, not the regional one, which was increasingly important during the Middle Woodland. As agricultural economies developed, local institutions became more central to survival. At the same time, the vestiges of the larger scale networks that had been crucial to centrifugal social organization remained relevant through the exchange of ideologically charged materials and symbols.

Another more recent model of community organization is useful for understanding the tension between bourgeoning local institutions and the maintenance of regional connections that seems to characterize the Middle Woodland period throughout eastern North America. Ruby et al. (2006) take examples from the Wabash River Valley, the Scioto-Paint Creek (southern Ohio) heartland, and the Lower Illinois River Valley to examine Middle Woodland communities at different scales. They point to three different types of communities discernible in the archaeological record: (1) a residential community, defined by coresidence and daily face-to-face interaction; (2) a sustainable community, defined by the pool of potential marriageable mates, estimated at a minimum of 500 individuals (Wobst 1974); and (3) symbolic communities, where membership is expressed through symbols that transcend physical or familial closeness (Ruby et al 2006: 123). The expression of these three levels of community is variable between the three areas discussed, but the scheme applies to all. I now turn to the expression of community organization specific to the Lower Illinois River Valley.

Middle Woodland Subsistence and Settlement in the Lower Illinois Valley

During the 1960s, research in the Lower Illinois River Valley came to exemplify processual approaches to Hopewell studies. Projects focused on paleoenvironmental reconstruction and settlement structure (Asch 1976; Asch et al. 1979; Farnsworth and Asch 1986; Struever 1964, 1968; Zawacki and Hausfater 1969), while mortuary studies began to explore population-wide paleopathology and biological distance (Braun 1979; Buikstra 1976, 1979). Over the past 30 years, the resolution of data describing the Middle Woodland period and its Hopewellian characteristics in the LIV has drastically improved, allowing researchers to place new data in the context of a recreated landscape that is continually coming into clearer focus.

The Lower Illinois Valley is defined as the approximately 115 km of river between Meredosia, Illinois, where the river turns south, and its confluence with the Mississippi at Grafton, Illinois. In its lower reaches, the gradient of the valley is dramatically reduced, exhibiting almost no relief and sometimes descending less than 2 cm per km. The floodplain is quite narrow, averaging 5.6 km wide, and is bounded by limestone bluffs mantled in Pleistocene loess deposits (Fig. 3). The uplands are deeply incised by tributary streams, which sometimes flow through smaller floodplains of their own. Where these streams enter the main floodplain, they form alluvial fans at the foot of the bluffs. Before the construction of modern flood management infrastructure, the river and its tributaries, as they flowed through the nearly flat floodplain, were bounded by natural levees. Annually, the river flooded its levees and formed backwater lakes between the levees and the bluffs. This flood regime was established by 6000 cal. BP (Wiant et al. 2009: 233).

Several scholars have attempted to reconstruct the ecology of the valley (Asch and Asch 1985a, c; 1986; Zawacki and Hausfater 1969). These reconstructions have generally relied on the first United States Government Land Ordinance (GLO) conducted in 1819, when the valley was just beginning to be settled by Euro-American farmers. Recognizing that the Land Surveys do not offer a perfect reconstruction of Middle Woodland ecology, these data do allow us to appreciate the elaborate mosaic of ecotones that characterized the valley before modern agriculture. The bottomland was largely covered by floodplain prairies, which dominated the poorly drained terraces flanking the river. Strips of floodplain forest clustered along the river and its major tributaries. The composition of floodplain plant communities was primarily determined by elevation, soil drainage, and soil aeration (Turner 1934). Mesic and dry-mesic hardwood forests or "barrens" of scattered oak and hazelnut thickets covered the bluff slopes. Tall grass prairie extended both east and west along the bluff tops, cut by creek bottom forests of oak and hickory. Fire and flooding both played a role in the creation of this ecosystem.

The landscape of the LIV was the site of some of the earliest experimentation with small seed crop production in the Eastern Woodlands (Fig. 4). Far from facilitating rapid socioeconomic change, early cultivation seems to have been gradually

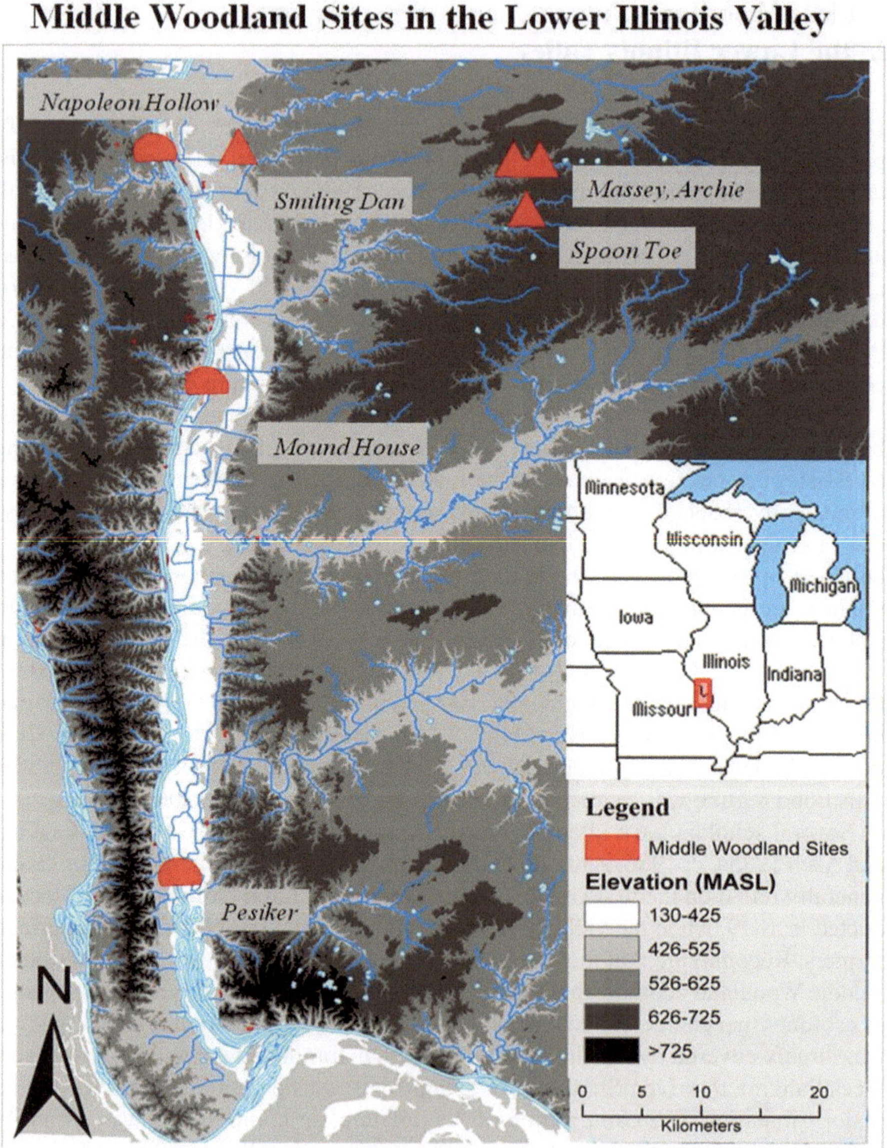

Middle Woodland Sites in the Lower Illinois Valley

Fig. 3 *Red* indicates Middle Woodland sites identified by the Illinois Archaeological Survey. *Semi-circle icons* are mound sites under analysis in this study; *triangles* are habitation sites. Permanent natural streams and lakes are pictured in *blue*. Note that the channelization of streams in the floodplain is of twentieth century origin

integrated into hunting and gathering societies. Possibly cultivation was made possible by increased sedentism, which in turn was facilitated by increasing reliance on floodplain aquatic resources. Backwater lakes were critical to the inhabitants of the LIV from c. 6000 B.P., when the modern flood regime emerged, through

Latin Name	Common Name	Earliest Archaeological Appearance	Modern Habitat	Native to W. Illinois?	Source
Ambrosia trifada	Giant Ragweed	7000 B.P.	Moist to mesic soils, alluvium, rich waste places	Yes	Asch and Asch 1985d:161, Cowen 1985:214, Zawacki and Hausfater 1969:51
Chenopodium sp.	goosefoot	8500 B.P.	Understory of willow river margin stands, floodplain sand banks, upland (terrace) clearings	Yes	Asch and Asch 1985d:171; Munson 1984:383; Smith 1992a:173
Cucurbita pepo	Squash	7000 B.P.	Sand bars, river banks, dumpheaps	No?	Asch and Asch 1985d:153-4, Fritz 2000:230
Helianthus annuus	Sunflower	7800 B.P.	Full sun, plains or bottomlands; will tolerate dry to moist soils	No	Asch and Asch 1985d:164-6
Hordeum pusillum	Little Barley	2770 B.P.	Along roads, pastures	Yes	Asch and Asch 1985d:190-3
Iva annua	Sumpweed/ Marsh elder	7320 B.P.	Open or disturbed riverbanks and lakeshores, wet soils	Yes	Asch and Asch 1985d:159-60
Phalaris caroliniana	Maygrass	2000 B.P.	Disturbed habitats, along roads and railroads, sandy soils, fallow fields but not recently plowed ones	No	Asch and Asch 1985d:188-90, Fritz 2011)
Polyganum erectum	Knotweed	5000 B.P.	Exclusively human disturbed areas: roadsides, paths, pastures; packed soils	Yes	Asch and Asch 1985d:184-5; Munson 1984:384

Fig. 4 Table showing the earliest appearance of plants at sites in the Lower Illinois Valley that would eventually be cultivated or domesticated

the Middle Woodland period. They concentrated aquatic resources throughout the warm season by trapping populations of fish and harboring abundant mussels, turtles, and water fowl.

These concentrated, high-yield resources enticed human populations to establish larger settlements at the margins of the main valley by the Late Archaic (Wiant et al. 2009). Edible seeds that would later be cultivated more intensively are commonly recovered from Late Archaic sites, but do not become abundant and ubiquitous until the Middle Woodland period.

The Eastern Agricultural Complex

Middle Woodland subsistence in the LIV is characterized by the emergence of the Eastern Agricultural Complex (EAC). The EAC is regionally variable, but in the LIV its most important components were little barley (*Hordeum pusillum*), maygrass (*Phalaris caroliniana*), erect knotweed (*Polygonum erectum*), and goosefoot (*Chenopodium berlandieri*). To avoid semantic confusion, I refer to the subsistence system of the LIV during the Middle Woodland as *agriculture*, because it constitutes one expression of the Eastern *Agricultural* Complex. Researchers draw various lines between the practices of cultivation, horticulture, and agriculture. For example, Scarry (1993: 7) defines agriculture as "large scale crop production, that is, where plots of land were planted on a long term basis," with fertility maintained by flooding or by short fallows. In her scheme, horticulture is characterized by smaller plots and longer fallows. This definition makes sense for the classification of modern subsistence systems, or well-recorded historical ones. But we do not know the size of Middle. Woodland plots, their location, nor the timing of their fallows. Rather than defining terms based on unknown parameters, it seems reasonable to refer to Middle Woodland plant husbandry as agriculture in the sense of *crop production*, on the strength of the evidence that *is* available: thousands of EAC crop seeds, processing tools, storage pits, and cookwares from the habitation areas of the LIV (Fritz 1990). To reiterate, agriculture is here defined exclusively on the basis of *its product*s (crops and attendant material culture) since very little is known about the *process* (fallows, field size, weeding, etc.)

The plants that make up the EAC are best conceptualized on what Bruce Smith (1992a) refers to as a continuum of human-plant relationships, following the scheme developed by Harlan and de Wet (1965). On one end of the continuum lie wild plants, some of which are useful to humans and some of which are not. In the middle of the continuum are weedy plants, which are consistently able to dominate human-disturbed environments. Humans may either attempt to eradicate these plants, tolerate their presence, or find them useful and encourage their growth. The latter is the mechanism by which *weedy* plants may become *cultivated* plants (Gremillion 1993). The crucial difference between weedy "camp followers" and cultivated plants is human propagation, or *the saving and sowing of seeds*. Depending on the plant, its use, and the intensity of human exploitation, cultivation *may* lead to domestication. Crucially for the case of the Eastern Woodlands, cultivated plants are not necessarily domesticates, but cultivation is the process by which wild cultivated plants may become domesticated. Finally, domesticates exhibit consistent genetic and/or morphological variation from their wild relatives; some are completely dependent on humans for their propagation.

For the crops discussed in detail here, there is no genetic evidence for domestication. For other domesticated plants, the DNA of modern plants can be compared to that of their wild relatives and, when it is preserved, to ancient DNA from older varieties of the same crop. In the Eastern Woodlands, the only native crop that is still cultivated is sunflower, which has been the target of several genetic studies investigating both the geographic origin of the domesticate and the genetic loci that led to morphological changes (Blackman et al. 2011; Burke et al. 2002; Wills and Burke 2006). Similar studies would be far more complicated for the other members of the EAC because the cultivated (and possibly domesticated) varieties are extinct, necessitating the extraction of ancient DNA for any comparison to wild plants. In the absence of such data, any case for domestication must be made on the basis of morphological changes alone.

Demonstrating domestication archaeologically is rarely straightforward because many of the changes that commonly accompany domestication are not visible in assemblages of individual carbonized seeds. The lack of morphological markers in non-domesticated cultivated plants necessitates building a case for cultivation based on several lines of evidence. Smith (1992b: 107–114) and Asch and Asch (1985d) have established criteria for recognizing cultivated plants and domesticates. I will briefly review the cases for little barley, maygrass, erect knotweed, and goosefoot using their criteria as a guide. It should be noted that there are several other plants whose status as indigenous eastern North American domesticates has been established based on morphological changes, including sunflower (*Helianthus annuus* var. *macrocarpus*), sumpweed (*Iva annua* var. *macrocarpa*), and squash (*Cucurbita pepo* ssp. *ovifera* var. *ovifera*), but none of these have been identified at Mound House (Smith 1992a, b; Yarnell 1972).

Little Barley (*Hordeum pusillum*). Limited morphological analyses of little barley caryopses (one-seeded fruit, typical of grasses) have shown that archaeological specimens fall within the same size range as modern experimentally charred caryopses (Asch and Asch 1985a: 370). While for other crops (goosefoot, maygrass) desiccated caches have given analysts the opportunity to study other possible metrics related to domestication, entire little barley plants in storage contexts have not been found. Thus, there is no evidence that Middle Woodland little barley was domesticated. In the southwestern United States, however, Bohrer (1991) has suggested that little barley was domesticated by Hohokam farmers of central Arizona.

Despite its wild morphology, researchers are generally in agreement that little barley was cultivated prehistorically. Range extension is one common argument supporting the cultivation of plants that show no morphological change, but the archaeological specimens from the LIV fall within the modern range of wild little barley. Nor has little barley been identified in human paleofeces, as have other members of the EAC. Absent these indicators, the case for cultivated little barley rests on several interrelated pieces of circumstantial evidence. Little barley today is a common plant in human-disturbed habitats, fitting Smith's (1992b: 107) criteria for "modern weed analogues": "based on its present day abundance in disturbed soil situations…[it was] also quite likely a weed within human-made habitats" in the past. Yet little barley is rare or absent at Archaic sites within its natural range, suggesting that its abundance at Middle Woodland sites cannot be explained exclusively as a result of a weedy tendency to grow in human habitations. Little barley seeds are found in the same contexts and abundance as other members of the EAC whose status as cultivars is supported by stronger lines of evidence. Its economic importance is attested to by an archaeological presence that spans North American from Arizona to Illinois (Asch and Asch 1985a: 372).

Maygrass (Phalaris caroliniana). Like little barley, there is no evidence that maygrass was domesticated, but its status as a cultivated plant is unequivocal because it is found in abundance at archaeological sites far to the north of its present wild range, which extends only to the Missouri Bootheel in the Mississippi River Valley (Fritz 2010). The ubiquitous and abundant maygrass at the sites in the LIV discussed here is thus far outside of its natural range and must have been planted by humans. Unlike little barley, maygrass is not particularly characterized as weedy, instead preferring open but not recently disturbed habitats such as fallow fields and roadsides. Producing maygrass would have required locating or creating forest openings, but maygrass is not a classic "dump heap colonizer" (Anderson 1952; Fritz 2010). In addition to its appearance outside of its natural range, its presence in human paleofeces attests to the dietary importance of maygrass to prehistoric people (Yarnell 1969). The presence of intact plants lacking only the inflorescences at rock shelters in Kentucky suggests that maygrass was harvested by uprooting the entire plant, in which case human propagation would have been absolutely necessary to the continuation of stands (Cowan 1985: 213).

Erect Knotweed (Polygonum erectum). Morphological variation among erect knotweed achenes (one-seeded fruit consisting of seed encased in a thin pericarp) complicates attempts to establish criteria for domestication. Erect knotweed achenes are dimorphic: one type is slightly shorter and wider with a thick reticulate pericarp; the other is longer and narrower with a smooth, thin pericarp. Both types occur in the same seed head of wild plants, but the smooth morph increases proportionally late in the season. Both morphs commonly appear in Middle Woodland assemblages from the LIV (Asch and Asch 1985b: 184). By the Mississippian period, it appears that a larger version of the smooth morph predominated among cultivated varieties in the LIV and in the Ozarks (Hill Creek and Whitney Bluff rockshelter sites, respectively), an observation that deserves much more extensive study and may indicate nascent domestication (Lopinot et al. 1991: 5). The thinner pericarp and larger size of the smooth morph are both signs of the "domestication syndrome" first described by De Wet and Harlan (1975): the thinner pericarp reduces the period of dormancy, and the larger fruit size provides more nutritive material and thus faster growing seedlings (Asch and Asch 1985b: 140–141; Lopinot et al. 1991). However, Lopinot et al. (1991: 2) point out that there have been no systematic studies of achene morphology in wild plants that take into account seasonal variation and consistently report size, shape, and pericarp texture in comparison to archaeological collections, leaving the possibility of Mississippian domesticated erect knotweed an open question.

     With the morphological case for domestication shaky at best, the argument for erect knotweed as a cultivated plant rests on similar circumstantial arguments as that for little barley. The natural range of wild erect knotweed encompasses the regions where it is most often recovered archaeologically: western Illinois and the American Bottom. However, it is recovered consistently and is often the most abundant plant type at Middle Woodland sites in these regions and is found in similar contexts as other crop seeds (Asch and Asch 1985d: 183–186). Asch and Asch (1985d: 185–186) have noted that today erect knotweed does not occur in stands large or dense enough to support productive harvests from wild stands, arguing that cultivation would have been necessary to obtain harvests sizeable enough to create the archaeological remains recovered. At later Mississippian sites in the LIV and American Bottom, masses of erect knotweed have been recovered from storage pits, lending further support to its classification as a cultivated plant and possible domesticate (Lopinot et al. 1991).

Goosefoot (*Chenopodium berlandieri and C. berlandieri* ssp. *jonesianum*). Goosefoot is the only member of the EAC found at Mound House for which there is unequivocal proof of domestication. It is important to note at the outset that while goosefoot was domesticated prehistorically, not *all* archaeological goosefoot is domesticated; both weedy and wild seeds are also present in most assemblages (Gremillion 1993). A preliminary analysis of the status of the Mound House goosefoot is provided below. The criteria for the native eastern North American domesticated subspecies of goosefoot, *Chenopodium berlandieri* ssp. *jonesianum*, were codified by Smith and Funk (1985) and based largely on the work of Hugh Wilson (1981; Wilson and Heiser 1979). The domesticated eastern North American variety is characterized by five specific morphological traits that indicate the domestication syndrome: (1) larger infructescences concentrated at the ends of branches; (2) loss of seed shatter mechanisms; (3) simultaneous flowering and uniform maturation of fruits; (4) larger perisperm (nutritive content of the seed), expressed by a change in seed shape from biconvex to truncate margins and the resultant increase in volume; and (5) reduction in the thickness of the outer epiderm (testa) (Smith 1992b: 110–115).

For carbonized collections of single seeds (as opposed to more rare whole plants preserved in rock shelters), only two of these criteria are useful. The thickness of wild *C. berlandieri* testas are 40 µm, while those from large collections of *ssp. jonesianum* are 20 µm (Smith and Yarnell 2009: 6563–6564). Ideally, testa thickness is measured using a scanning electron microscope. However, the large difference between the wild and domesticated testa thicknesses allows for a preliminary characterization using a standard optical microscope. Likewise, the shape of the seed margins is readily observable at low magnification. Wild seeds and weedy seeds have biconvex, rounded, or equatorial-banded margins, whereas ssp. *jonesianum* has a truncate margin, especially opposite the "beak" (embryonic root) (Smith 1992b: 111–112). In addition to the morphological changes in goosefoot, the other arguments for cultivation common to all members of the EAC also apply. Goosefoot seeds are ubiquitous and abundant in Middle Woodland contexts, they are present in human paleofeces, and they are amenable to cultivation because of their natural preference for openings and disturbed ground (Asch and Asch 1985d: 171–183; Yarnell 1969).

The proceeding discussion has attempted to describe *what* Middle Woodland people cultivated. A study of settlement structure and site function is necessary to explore *how* they organized production, exchange, and consumption of cultivated plants.

Settlement Structure and Site Function

The Middle Woodland mounds of the LIV were first explored as an analytical unit by Stuart Struever (1968). His investigations were driven by the hypothesis that changes in subsistence were tied to the development of the Hopewell phenomenon. He divided Middle Woodland sites in the LIV into five categories including base camps, summer agricultural camps, mound groups (floodplain mounds), burial sites (bluff top mounds), and one "regional exchange center"—Mound House. Although he never excavated Mound House, surface scatter of Hopewell Interaction Sphere materials (obsidian, mica, copper, effigy pipes, etc,) led him to believe that this site held a unique position in the settlement structure of the LIV, despite the fact that it is one of several floodplain mound complexes in the valley (1968).

Later surveys and reevaluation of the early material have shown that Mound House is unexceptional in terms of extra-regional exchange items (Farnsworth 1990), and Struever de-emphasized Mound House as a locus of interregional exchange in his later interpretations, instead proposing that the Golden Eagle Mound Group, at the confluence of the Mississippi and Illinois, may have served this purpose (Struever and Houart 1972). Golden Eagle has not been the subject of modern archaeological investigation, so its role in exchange and ritual remains unclear. Meanwhile, excavations at Mound House have demonstrated that it is unlikely to have functioned as a central place for interregional exchange. Lithic artifacts recovered from the site are extremely diverse in terms of function, but lithic raw materials are not particularly exotic. Likewise, the Mound House ceramic assemblage does not contain unusually large amounts of Hopewell ceramics or items of nonlocal manufacture (Buikstra et al. 1998: 27, 32). While Mound House was evidently a ritual center, it does not seem to have functioned as a locus of interregional exchange any more so than some residential sites. The hamlet of Smiling Dan, for example, has a similar ceramic assemblage and number of Hopewell Interaction Sphere items and a greater diversity of lithic raw materials, in terms of source material (Buikstra et al. 1998: 46–59). Current thinking on the settlement structure of the LIV varies considerably from Struever's original model. The following discussion closely follows the interpretation provided by Ruby et al. (2006) in their comparative study of Middle Woodland community organization, with supplementary evidence from earlier investigations.

Bluff-Base Hamlets. As in Ohio and the American Bottom, systematic surveys have revealed that Middle Woodland people did not live in large, nucleated villages. Instead, they lived in hamlets or individual homesteads. In the LIV, surveys have shown that the best predictor of hamlet location is proximity to a non-stagnant water source, usually either where the Illinois River flows near the bluff base or where a tributary stream enters the main valley (Asch et al. 1979). These habitations are thus best characterized as bluff-base hamlets. Bluff-base hamlets consist of two to four houses, usually sub-rectangular and 6–8 m on a side, surrounded by shallow pits and sheet midden deposits. Large dumps in ditches or along terrace edges are common and can be 1–2 m thick.

The bluff-base hamlets of the LIV (as well as its mound centers) share a ceramic complex referred to as the Mound House phase. The Mound House phase consists of a mix of Havana/ Hopewell and Pike/Baehr types. These are paired complexes of utilitarian and fancy wares. The Havana/Hopewell complex seems to be derived from Central Illinois Valley pottery and forms the basis for argument that settlers from the central valley moved into the LIV at the beginning of the Middle Woodland period (Farnsworth and Asch 1986). Pike/Baehr may be either a later iteration of this style or a contemporary local tradition (Buikstra et al. 1998). Ruby and colleagues (2006: 132) define the Mound House phase as "an occupation of the LIV by peoples of the Havana (Hopewell) tradition," between 50 BCE and 250 CE. This type of definition references exchange-based conceptions of Hopewell: that is, the Middle Woodland occupation of the lower valley is defined by its material ties to other regions. Hopewell and Baehr vessels, which are small, finely made, and elaborately decorated pots, are concentrated in the Middle and Lower Illinois Valley and are earlier in the middle valley. This situation has led researchers to conceptualize these areas of centers of production, both physically (of pots) and ideologically (of their attendant symbolic systems and related traditions) (Fie 2006: 432).

Recent compositional analysis has confirmed that both Hopewell and Baehr ceramics were produced in the LIV (Fie 2006: 437–438). Although these symbolically charged vessels were often deposited in burials, they were evidently produced and/or used in hamlets as well; Hopewell vessels are better represented at Smiling Dan than they are at Mound House. Ceramics produced in southern Illinois and the Central Illinois Valley are also occasionally found at hamlets, as are locally made pots that bear stylistic similarities to extra-regional traditions (Fie 2006). This complex situation suggests the occasional movement of both materials and people into the hamlets of the LIV, as well as the production of Havana Hopewell symbolic culture at habitation sites with no evidence of mortuary ritual.

The most fully excavated and published bluff-base hamlet is Smiling Dan, which was occupied by one or two households during the Mound House phase. The scale of the middens and pits and the composition of the botanical, faunal, and tool assemblages make it likely that bluff-base hamlets like Smiling Dan were occupied year round, although individuals evidently moved around the landscape frequently to harvest plants, hunt, and engage in rituals and trade (Ruby et al. 2006: 134; Stafford and Sant 1985). Hamlets were not evenly distributed throughout the valley. Although only a handful have been excavated (including Smiling Dan, Macoupin, Apple Creek, and the Gardens of Kampsville), surveys show that bluff-base hamlets are clustered in groups of two to five separated by 0.8–1.6 km, with longer distances between one cluster and the next (Ruby et al. 2006: 134). While clusters of hamlets cannot be characterized as villages, inhabitants of different hamlets in the same cluster could have communicated with each other on a daily basis, meeting the basic criteria of residential community.

Upland Hamlets. The hamlets and single households of the upland tributary valleys are somewhat more enigmatic. Four of these uplands sites have been excavated (Massey, Archie, Missed Point, and Spoon Toe); all are located in the eastern tributary valleys. These consist of maximally two households, which were probably not occupied simultaneously (Asch and Asch 1985c; Calentine 2005). Aside from being slightly smaller than bluff-base hamlets, they also share a distinct material culture. All four have been assigned to a separate ceramic phase than the sites in the main valley, called Massey phase after its type site. This style of pottery is distinct from the Mound House phase ceramic assemblage characteristic of bluff-base hamlets and mound complexes in the LIV. The Massey phase sites in the uplands are con-temporaneous with the later part of the Mound House phase: the first two centuries CE. However, stylistically they are more similar to Crab Orchard type pottery from southern Illinois (Farnsworth and Koski 1985: 225). While the ceramics at the Massey phase sites resemble Crab Orchard pottery, they were manufactured with local raw materials (Calentine 2005: 89).

It is most likely that at least the potters at the Massey phase site were settlers from southern Illinois or their descendents. Trade connections at Massey phase sites are with the main valley, not southern Illinois, so it is unlikely that locals were inspired by trade goods to try new styles (Farnsworth and Koski 1985). Hopewell and Baehr style bowls, Norton and Gibson style projectile points, and lamellar blades have been found at Massey phase sites. All are artifacts common at main valley sites and diagnostic of the Mound House phase. Furthermore, both the botanical and lithic assemblages from Massey and Archie, two upland homesteads, show that these sites were minimally occupied during the summer and fall, and extensive middens and pits make year-round occupation a possibility. Radiocarbon dates demonstrate that these two households were occupied for a minimum of 100 years (Farnsworth and Koski 1985). These do not seem to be the campsites of visitors to the valley, who in any case would be unlikely to produce copious amounts of pottery while traveling far from home. It is possible that some of the inhabitants of the Massey phase sites were locals of the LIV, but that the potters, perhaps women, were from farther afield and maintained their traditional ceramic style. In any case these sites provide evi-dence of population movement into the valley from the south, just as the Havana ceramics of the main valley provide evidence of settlement from the north.

Bluff-Top Mound Complexes. In addition to bluff-base and upland hamlets, mound complexes in the valley can be divided into two broad categories: bluff-top and floodplain. Bluff-top mounds are ubiquitous in the LIV and served as cemeteries and centers for ongoing mortuary ritual. With the exception of the anomalous loaf-shaped Naples-Russell Mound 8, these are conical and are arranged in linear groups overlooking the valley. Burial rituals were complex and variable, but generally Middle Woodland burials were serial affairs in which a body was interred in a central, log-lined tomb and later redeposited around and within the mound as disarticulated or bundled remains. Burials are often accompanied by exotic Hopewell exchange items, such as mica, copper, pipes, galena, and marine shells (Buikstra 1976).

People did not live amongst bluff-top mounds, but Napoleon Hollow is a habita-tion site closely associated with the Elizabeth mound group. It is located at the foot of the bluff directly below the mounds. Its excavators point to a near lack of plant processing and hunting tools and a homogenous faunal assemblage to argue that residents were being provisioned and were primarily engaged in ritual activity at the site. They propose that Napoleon Hollow functioned as a "ritual camp" where little or no hunting and agriculture took place (McGimsey and Wiant 1986: 540). Alternatively, they may have provisioned themselves, that is, transported stored food to the site from their homes. Whether or not visitors to Napoleon Hollow were able to com-mand provisioning is an open question, but the patterning of the refuse they left does indicate that while they were at the site, they were not engaging in normal subsistence-related activities.

Bluff-top mounds have been characterized as community cemeteries associated with discrete territories, used by a single community over time rather than by several at the same time (Charles 1995; Konigsberg and Buikstra 1995). The position and relative age of the bluff-top mounds has been used to document the resettlement of the valley after a near abandonment during the Early Woodland period. Charles' classic model postulates that as immigrants and growing kin groups filled in the LIV from north to south, they expressed their corporate identities through an elaboration of mortuary ritual. Charles (1995: 90) writes: "The trade partnerships and the funerals, which were pan-local affairs, maintained ties to other communities that could be tapped" in the event of local resource shortfalls (1995: 90). He echoes Ford's idea of centrifugal community organization and the sites of formal visitation that stabilize them. Recently, direct dating of burials and settlement sites in the LIV has supported the scenario of a north to south trend in mound building beginning in the first century BCE (King et al. 2011).

Floodplain Mound Complexes. Although dates from floodplain mound complexes are still sparse, it seems that they were built later than the earliest community cemeteries and habitation sites in the north section of the valley (King et al. 2011). If this chronology is accurate, it indicates an elaboration of local institutions for communication and exchange over time, as predicted by Ford's model. Floodplain mounds were also mortuary facilities, but they differed from bluff-top mounds in several respects. In general, they are larger and contain more burials per mound, but fewer burials overall, than bluff-top mounds. The central tombs also tend to be larger. Importantly, burials at floodplain mounds are almost exclusively adult males, while populations at bluff-top mounds include roughly equal numbers of males and females and a cross section of population by age at death. Men buried in floodplain mounds are also, on average, taller than their contemporaries interred on the bluff tops. There is no evidence that the men in the floodplain mounds were all members of the same family and they are no more closely related to each other than to other inhabitants of the valley (Buikstra 1976: 40–42). This evidence suggests that these men held a special position in society, but that their status was not based on membership in an elite kin group.

The nature of habitation at Mound House is still very poorly understood, but excavations at Peisker, a similar floodplain mound center in the LIV, provide some context. Based on faunal, lithic, and ceramic remains, Peisker seems to have functioned as both a center for mortuary ritual and a habitation occupied either in the spring or autumn (Staab 1984). Middle Woodland burial practices and other ritual activities at mound groups in the LIV were iterative and required that people return regularly to the mound sites to build and extend mounds and to process and reinter the dead. Floodplain mounds were places that people from different residential communities returned to at fixed times to participate in a larger symbolic community. Napoleon Hollow, Peisker, and Mound House provide insight into the range of activities carried out at areas of formal visitation, including plant use.

Summary of LIV Settlement Structure. Struever's schematic functional characterizations of LIV sites along economic lines have not been borne out. Floodplain mound sites were not necessarily centers for interregional trade and were evidently inhabited seasonally. Non-mound habitation sites are more accurately grouped by physiographic location than by economic specialization. The LIV is landscape of relatively drastic ecological variability over very short distances, a fact which must have driven human decisions about where to live. New evidence has done little to repudiate the low population density estimates proposed in the 1970s, which were based on the typically small settlement size and known burial populations (Asch 1976).

Faunal remains leave no doubt that, even as agriculture was developing, the abundant resources of the Illinois River were still at the heart of Middle Woodland subsistence. Several researchers have argued convincingly that floods and river fauna were also central to ritual life and symbolism (Charles et al. 2004; Van Nest 2006). But the floodplain was not conducive to year-round habitation without the extensive flood management infrastructure that exists today; much of the main valley would have been under water throughout the spring and sometimes into the summer. As a result, habitation sites in the valley tend to be located on the higher terraces at the foot of the bluff, where tributary creeks run into the floodplain. In this analysis, Smiling Dan will represent this type of site: a habitation consisting of more than one residence and storage features at the margin of the main valley (Stafford and Sant 1985).

Smaller but structurally similar sites are scattered along the narrow floodplains of the upland creeks. Contrary to Struever's expectation that early agricultural was tied to the major river valleys, paleoethnobotanical investigations have shown that upland sites were more likely to be specialized agricultural camps than hunting bases, a conclusion that has recently been reiterated (Asch and Asch 1985c; Calentine 2005). Given the risks associated with farming in an unmodified floodplain where summer inundations are not uncommon, it is not surprising that occupants of upland habitations participated in seed crop harvesting and processing, although these activities took place at bluff-base habitations as well. Excavated examples of this type of site may also provide evidence of communities with more ties to southern Illinois than the population of the main valley, based on the ceramics they created. Massey, Archie, and Spoon Toe will serve as examples of this type of site. They are small, isolated dwellings surrounded by shallow pits near the floodplain of tributary creeks (Calentine 2005; Asch and Asch1985c).

Detailed comparisons of the patterning of different material classes between sites in the LIV are common and have greatly improved our understanding of how Middle Woodland society was structured in this region. This analysis will proceed with three fundamental assumptions: (1) plant-related activity varied between sites in the valley, (2) variability in behavior can be seen in the patterning of botanical refuse, and (3) variability can help us better understand the broader Middle Woodland trend of agricultural intensification. Mound House, Smiling Dan, and the Massey phase sites represent habitation in the three major physiographic regions of the valley, the floodplain, bluff-base, and upland creeks. Botanical assemblages from Peisker and Napoleon Hollow are the residues of formal visitations to floodplain and bluff-base mound complexes, respectively. A comparison between these two sites and Mound House may show how (or if) subsistence-related activity at mound complexes varied according to their ritual function. Methodologically, these botanical assemblages lend themselves to comparative study because all but Peisker and Spoon Toe were analyzed using the same methods.

Plant Use at Mound House

Previous Investigations

Struever (1968) was the first to execute a formal surface survey of Mound House, in which he also reported the results of several previous attempts. He based his interpretation of Mound House as an interregional interaction center on the results of these surveys. James Bellis was the next to survey the site. His investigations defined Mound House as encompassing 5 ha of scattered cultural debris associated with two extant mounds, revealed three areas of concentrated remains, and turned up artifacts dating to the Archaic, Middle and Late Woodland, and Mississippian Periods, with Middle Woodland artifacts constituting the majority (Buikstra et al. 1998: 15)

The first excavations at Mound House took place in 1986 when the Center for American Archaeology conducted a field school at the site. Sixteen 2-by-2 m test units were excavated (12 were terminated at the base of the plow zone), and a systematic survey of 463 5-by-5 m units was conducted. Features including pits and post-molds were identified in excavation units. Density plots from this unpublished report are reproduced by Buikstra et al. (1998: 19–26). They reveal an extremely dense debris scatter north of the mound, where current excavations are focused, as well as a southern area of lesser concentration between Mounds 1 and 2 (Buikstra et al. 1998).

Interestingly, Pike-type ceramics are concentrated in the southern cluster and Havana-type sherds are more common in the northern cluster (Buikstra et al. 1999). Researchers disagree on whether Havana is an earlier tradition than Pike, represents a different ethnic group, or is merely functionally distinct. It is therefore possible that the northern activity area from which the assemblage under analysis was taken is (1) representative of a slightly earlier occupation than the southern (mid-mound) area, (2) the activity area of one of two ethnic groups who visited Mound House, or (3) an area where activities functionally associated with Havana ceramics were carried out. Buikstra (1998: 21–22) notes that items of nonlocal origin and lamellar blades are both significantly correlated with Havana sherds and in general the preponderance of blades and cores at the site suggests that it may have been a center of blade production. No matter which of the three possibilities explains the ceramic patterning, Havana pottery was evidently used and discarded in the same areas where food preparation, and potentially exchange, took place and from which the plant remains in this analysis were recovered. A final item of interest from these investigations directly pertains to plant use at the site: hoes, manos, and metates are relatively rare at Mound House in comparison to other sites in the valley (Buikstra et al. 1998: 21).

In an attempt to reevaluate the function of floodplain mound centers, excavations began at Mound House in 1990 under the auspices of University of Chicago and the Center for American Archaeology. These excavations were focused on "determining the cultural context of mound building" (Buikstra et al. 1998: 28). Toward this end, a 6-by-7 m "macroblock" was excavated down to pre-mound surfaces. Nine shallow pits and 112 post-molds in association with early Havana and Hopewell ceramics were uncovered below the southern slope of Mound 1. These sub-mound features have returned dates ranging from cal. 90 BCE to 40 CE ±70 (1998: 37). The excavators interpret the sub-mound features as the residue of iterative ritual activities. A circular structure with a yellow sand floor surrounded by a wooden palisade was removed and rebuilt several times. After each episode of post removal, the old post holes were filled in with yellow sand. After several reuses, the circle was expanded so that the post holes now form three concentric circles. This palisade structure, and other similarly iterative features, indicates multiple episodes of use and construction, leading the excavators to suggest that Mound House was inhabited in accordance with a seasonal ceremonial calendar (Buikstra et al. 1998: 83–85).

The yellow sand layer and its associated structure are covered by a thick layer of redeposited midden, which forms the base of the mound. Dates from this fill range from cal. 10–150 CE ±70. The period of habitation that created the midden fill thus overlapped with or directly followed the earliest ritual construction at the site (the yellow sand structure). Two pit features (232 and 258) originate at the top of the midden layer and cut into the underlying yellow sand, indicating that the midden surface was occupied for a period of time before the mound was built on top of it (Buikstra et al. 1998: 98–99). The redeposited midden is overlain by a layer of laminated silts in one of the two trenches, possibly indicating an episode of flooding, followed by the final mound construction. The extant uppermost layer of the mound is nearly sterile (Buikstra et al. 1998: 35–77). Because no dates are yet available from the new excavations, it is impossible to say exactly how this sequence of events relates to the creation of Feature 379, the large midden-filled pit from which the new botanical assemblage was sampled, although the latter is contemporaneous with at least part of the mound-building sequence based on artifact and ceramic styles.

A paleoethnobotanical analysis of the charcoal from the sub-mound features and redeposited midden was carried out by Marjorie B. Schroeder (1998). Schroeder analyzed 22 samples or 398.25 l of sediment. Her samples were taken from a variety of contexts, the nature of which is sometimes unclear because only features are described in detail. However, comparing the square and level information with the published profiles, it seems that the majority of the samples (probably 16 of 22) were taken either from the redeposited midden layer at the base of the mound or from the two features that originate in it. The remaining six were taken from sub-mound shallow pits in or near the iterative ritual structure or palisade. Not knowing which, if either, of these contexts is contemporaneous with the Feature 379 samples, I will present my analysis separately and in comparison to Schroeder's findings.

If new dates make the temporal structure of the site clearer, the two analyses can easily be combined in future studies.

Geoarchaeological analysis of Mound 1 suggests that ritual activity at Mound House was related to flooding and world-renewal ceremonialism. Building off of earlier arguments that Middle Woodland ceremonialism is reflected in historical world-renewal ideologies among various American Indian groups (Hall 1979), this argument employs ethnographic accounts from historic tribes of the Plains to explain the use of inverted sod blocks as a building material in Mound 1 (Van Nest 2006). These accounts record rituals associated with the widespread American Indian creation myth in which a creature (variably a turtle, duck, muskrat, or woman) dives into the primordial waters and brings up a handful of mud, which becomes the first solid land (Van Nest 2006: 404–405). In the Blackfoot Sun Dance, blocks of sod symbolize this first primordial soil. The sod blocks used in construction of floodplain mounds may reference a progenitor of these myths, in which case the mound itself is symbolic of the first earth rising out of the primordial sea (Charles et al. 2004: 58–59).

The location of Mound House also strongly suggests this allusion, as it is located on a sand ridge that would have been surrounded by water during spring floods and beside a backwater lake that would have held water for most of every summer (Van Nest 2006). If the ethnographic connection to the earth diver myth is correct, this setting would have allowed a metaphorical reenactment of the creation of the earth. Van Nest argues that the banks of backwater lakes (on which Mound 1 is built) might have acted literally as stages for divers reenacting the myth at ritual gatherings (2006: 425). Proximity to backwater lakes would also have been most attractive in early to mid summer, when fish had become trapped and useful plants had begun to grow in the newly deposited silts.

From 2009 to 2011, excavations have focused on the area of concentrated debris and subsurface anomalies (identified by magnetomet survey) to the north of Mound 1. Analyses of other material classes from these excavations are not yet available, but the botanical remains can begin to offer additional insights into the nature and timing of habitation at Mound House.

Methods

Squares 609 and 613 are adjacent 1-by-2 m excavations units from which the samples under analysis were taken. They encompass the eastern end and center of Feature 379, a 1-by-3 m pit filled with highly concentrated mixed debris. Both were excavated in 10-cm arbitrary zones. The samples used for this analysis were primarily taken from Zones 4–7, although three samples from Zones 2 and 3, which contained far less plant material, were included for comparison. Zones were usually divided into two or more components on the basis of changes in soil color or texture and concentrations of artifacts and organic materials. Feature 379 is characterized by extremely complex stratigraphy and seems to have been filled during a long series of discrete depositional events over a period of time as yet undetermined. Excavators took at least one 10-l flotation sample from each component, unless the component itself was smaller than 10 l in which case the entire component was floated. In a few cases, flotation samples larger than 10 l were taken from components which appeared to be heavily organic. In addition to the plant remains reviewed below, 3,056 g of bone, 2,292 g of mussel shell, 19, 035 g of limestone (associated with earth ovens at contemporaneous sites), and 402 g of utilitarian ceramic sherds were recovered from square 609 alone (King et al. 2010). The original function of the pit is unknown, but all of the material recovered can be characterized as midden. There is no evidence of *in situ* burning.

Flotation was carried out by students and staff at the Center for American Archaeology in Kampsville, IL, using a SMAP-type flotation system. Heavy fractions are recovered from the inner flot tank, lined with 0.86mm mesh; the light fraction is collected in a cheese cloth. Both are air-dried and stored at Kampsville in plastic bags. The samples analyzed here are only a small fraction of the total available from Feature 379, to say nothing of the hundreds of samples from other off-mound features, which should be the subject of future study. Initially, I selected a random sample from each zone in Square 609 for sorting. In the lab, it quickly became clear that samples from Zones 4 and below were much richer in plant remains. This may be either an artifact of preservation or of changes in ancient deposition over time. In the interest of time, I selected additional samples from lower levels because they yielded more information per lab hour. I also made an attempt to analyze samples from different depositional events, as defined by the excavators in their floor maps and profiles. There is considerable variability in the contents of the samples, which supports the impression formed by the excavators, that the pit was filled in a series of discreet episodes rather than as the result of one event. No dates are yet available.

I analyzed 17 samples in the Paleoethnobotany Laboratory at Washington University between January 2011 and November 2011. I divided both light and heavy fractions using geological sieves (2 mm, 1.7 mm, 1 mm, 0.71 mm, 0.425 mm) and examined samples under a stereoscopic microscope at up to 40× magnification. For the light fractions, I sorted and weighed all material greater than 2 mm (bone, shell, sediment, wood charcoal, seeds, etc.), but only seeds and fragments of acorn shell were pulled from the less than 2 mm fraction, and counts were recorded. The procedure is to also pull fragments of squash rind and maize from this fraction, but none were found. For the heavy fractions, this procedure was slightly modified. Heavy fractions were extremely large due to the richness of this particular feature; they often contained large animal bones, snail and mussel shell, debitage, ceramics, and limestone. Sorting all material greater than 2 mm would have been very time-consuming and of limited relevance to this study, so I removed only charred plant material from heavy fractions and recorded weights and counts. Additionally, if the 0.71 mm fraction yielded no seeds (as was usually the case), then the 0.425 fraction was not scanned. For both heavy and light fractions, I did not examine the pan.

I made identifications to the lowest possible taxonomic unit in consultation with Gayle Fritz and with the help of the Washington University Paleoethnobotany Lab reference collection and texts (Martin 1995; Montgomery 1977; Steyermark 1981). In cases where identification of seed fragments was possible, both a total fragment count and Seed Number Estimates (SNEs) were recorded, but SNEs are used in all calculations of density, ubiquity, etc. I made no attempt to identify wood fragments in this analysis.

Results

Of the 17 samples analyzed, 100% contained identifiable plant remains. Figure 5 summarizes the density of wood, nutshell, total seeds, and members of the Eastern Agricultural Complex (in this case, *Chenopodium berlandieri*, *Hordeum pusillum*, *Phalaris caroliniana, and Polygonum erectum*) in the assemblage as a whole. A total of 24 taxa were identified, including nuts, edible seeds, and fruits; these are presented in Fig. 6. The appendix provides a full list of identifications, with weights and counts, by context.

All of the samples under analysis come from the same feature, but they show considerable variability through the profile. Figure 7 shows nutshell and wood density by weight and seed density by count, with the samples listed in order of descending depth. Anomalously dense samples stand out. For instance, sample 94 has a seed density that is more than twice as high as any other sample, with a total of 329 seeds. Of these, 192 are *Chenopodium berlandieri* and represent by far the largest cluster of crop seeds in the assemblage. Sample 57 contains almost 9 g of nutshell and roughly twice as much walnut shell as any other sample. Samples 94 and 57 may represent the residues of seed and nut processing, although they are not nearly as

Number of Samples	17
Liters of Sediment	139.5
Total Weight*	5925.21 g
Wood Density	.13 g/liter
Nutshell Density	.19 g/liter
Total Seed Density	6.46 /liter
Eastern Agricultural Complex Seed Density	4.06/liter

Fig. 5 Summary of botanical assemblage from Feature 379. *The heavy fractions of 2/17 samples were re-floated at Kampsville. The total weight before re-floating is not available

Latin Name	Common Name
Amaranthus sp.	Amaranth
Carya sp.	Hickory
Chenopodium berlandieri	Goosefoot
Corylus americana	Hazelnut
Diosopros virginiana	Persimmon
Echinocloa sp.	Barnyard Grass
Festucoid grass	Fescue tribe of the grass family
Galium sp.	Bedstraw
Hordeum pusillum	Little Barley
Julans nigra	Black Walnut
Nelumbo lutea	American Lotus
Panicoid grass	Panic Grass subfamily
Phalaris carolinian	Maygrass
Phytolacca americana	Pokeweed
Poaceae	Grass family
Polygonum erectum	Erect Knotweed
Polygonum sp.	Knotweed or Smartweed
Portulaca oleracea	Purslane
Quercus sp.	Oak
Rhus sp.	Sumac
Rubus sp.	Raspberry/Blackberry
Vaccinium sp.	Blueberry
Verbena sp.	Verbena/Vervain
Vitis spp.	Wild Grape

Fig. 6 List of taxa identified from Feature 379 botanical samples

dense or homogenous as some concentrations recovered from other sites in the valley. The nature of the Mound House samples, containing an assortment of seeds, nutshell, and wood fragments, suggests the residues from cooking fires where many types of useful plants would be mixed together in small amounts.

Nutshell. The assemblage contains the shells of five different taxa of edible nuts: hazelnut (*Corylus americana*), thick- and thin-shelled hickory (*Carya spp.*), oak/acorn (*Quercus sp.*), and black walnut (*Juglans nigra*). Figure 8 shows their relative densities by weight. The category Juglandaceae is made up of fragments that are not sufficiently well preserved to distinguish between walnut and hickory, both members of the family Juglandaceae. Similarly, "*Nutshell, indet.*" indicates fragments which could be either Juglandaceae or *Corylus americana* (generally very small fragments).

The wet floodplain forest that most likely surrounded Mound House was not as rich in nut-bearing trees as the upland associations available to bluff-base villages. However, these nuts were probably still available in the vicinity of Mound House. The western bluff is very close to Mound House, and hickory and walnut trees were

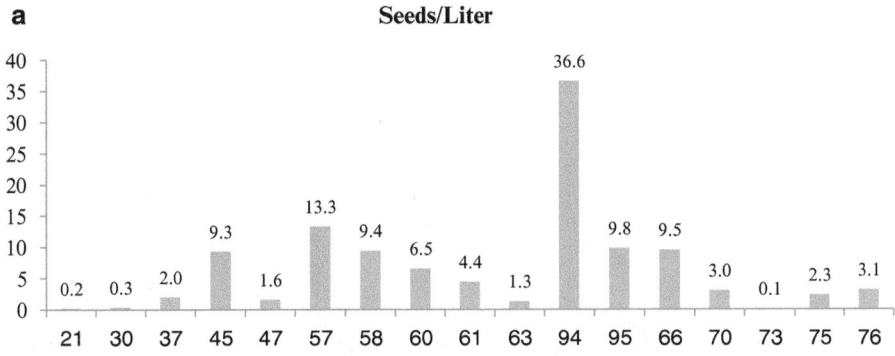

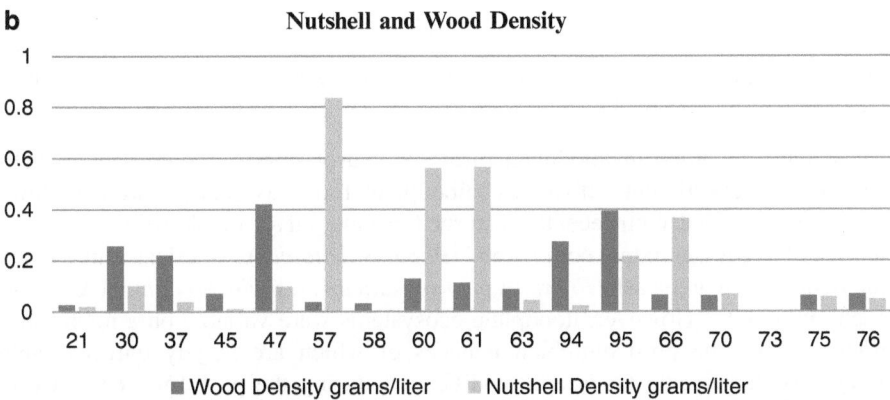

Fig. 7 Sample numbers listed on *X* axis in order of descending depth below surface. (**a**) Standardized density of seeds by sample. All whole seeds and weights of all fragments of wood charcoal and nutshell greater than 2 mm are included in calculations. (**b**) Nutshell and wood density by weight (g) by sample number. All fragments greater than 2 mm weighed

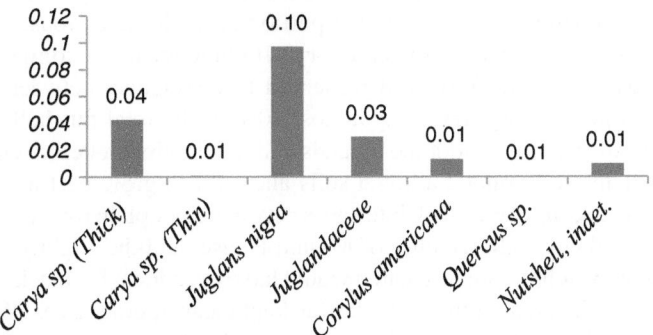

Fig. 8 Nutshell density by weight for all identifiable taxa

Nutshell, % of Total by Weight

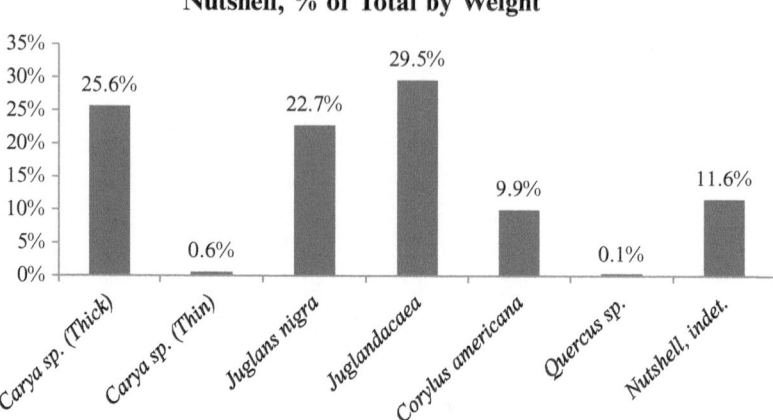

Fig. 9 The proportion of the total nutshell assemblage (fragments >2 mm) made up by each taxa

occasionally recorded in floodplain forests by Euro-American surveyors as well. The primary constituents of the wet floodplain forest association were willows (*Salix sp.*), especially on recently disturbed ground, silver maple (*Acer saccharinum*), and to a lesser extent cottonwood (*Populus deltoides*), American elm (*Ulmus americana*), and green ash (*Fraxinus pennsylvanica* var. *subintegerrima*) (Asch and Asch 1986: 449). However, floodplain ecosystems were variable on a microtopographic scale, the pre-Columbian nuances of which are largely unrecoverable because of large-scale landscape modification during the past 150 years. Mesic floodplain forests contained a greater diversity of species, including oaks, walnuts, and hickories, but were usually of very limited extent on the higher natural levees of tributary streams and the colluvial slopes of the main valley (Asch and Asch 1986: 449). The sand bank on which Mound House is located may have naturally supported a mesic floodplain forest, as would the terraces of the main valley, just across the river to the west.

Black walnut has the highest density by weight at Mound House and makes up the greatest single fraction of the total nutshell assemblage (Figs. 8 and 9). In fact, black walnut probably makes up a larger portion of the Mound House nutshell than is indicated by Fig. 9 because the category "Juglandaceae" is composed of fragments that are not sufficiently well preserved to distinguish between walnut and hickory, and this category makes up almost 30% of the total nutshell assemblage. Black walnut trees grew in both the uplands and the floodplain before clearance, but they prefer rich, well-drained alluvial soils and tend to grow best at forest edges. There may have been several such interfaces between floodplain forest and floodplain prairie or marshland in the vicinity of Mound House, and the slightly higher, well-drained soil on which the site is located would have been ideal for black walnut trees, even more so if the area around the site was kept clear of other trees. Walnuts were a food source for Woodland period communities, but they make up a small part of the nutshell assemblage at most sites in the LIV. It is possible that the black walnut shell at Mound House also had a technological or ritual use, as walnuts have been used by both American Indians and Euro-Americans as a pigment or for dying.

Thick-shelled hickory is the next most common type of nutshell at Mound House. It is less well represented than is usual for Middle Woodland sites in West Central Illinois. This may be because of the position of the site, relatively far away from the upland forests where hickory trees are so abundant. Because it is difficult to remove whole nutmeats from hickory shells, hickory nuts were probably used to make hickory oil or other processed foods, like the historically known *kenuchee* (hickory soup), processes that gradually separate nutmeats from hulls by boiling and pulverizing. If this was the case for Middle Woodland societies in the LIV, then hickory products may have been consumed at Mound House but processed there less often than at other sites.

Hazelnut shell is also present at Mound House but, again, in smaller amounts than is common for habitation sites in the LIV. It makes up only 6% of the nutshell assemblage by weight. Surveyors of the LIV in the nineteenth century noted hazels as common undergrowth shrubs, but they did not grow in the floodplain. They were common in oak-hickory forests with relatively open canopies but grew best as prairie thickets or at the interface between prairies and forests (Asch and Asch 1985a: 351). A recent synthesis of American Bottom prehistoric plant use noted a marked increase in the abundance of hazelnut shell at Middle Woodland sites (Simon and Parker 2006). This same pattern holds for the LIV.

There are several possible explanations for this trend. Hazels grow best at forest edges, so the establishment of small fields in the forest would have increased their habitat and created another resource for people to harvest in close proximity to their fields. However, if the intensity of land clearance was the only factor determining hazelnut frequency onsite, then incidence should steadily increase, or occasionally peak, from the Late Archaic (when cultivation of weedy seed crops began) until the Mississippian. This is not the case. The abundance of hazelnut during the Middle Woodland period could simply indicate a cultural preference which led people to cultivate hazel to increase yield during this time, but not during other periods (Simon and Parker 2006: 224–225).

There is a third possible explanation for the abundance of hazelnuts particularly in the Middle Woodland, which is that land clearance or forest thinning by fire was most intense during this period (Asch and Asch 1985a 352–355; Simon and Parker 2006). The relationship between hazel and fire, whether natural or human induced, is well established. For example, one study which compared undisturbed timber stands to recently burned stands in the Turtle Mountains of North Dakota found that the incidence of hazel in the burned stands had increased by nearly 25%. Although studying drastic changes in the overall plant community, the authors assert: "The most obvious influence of fire is the increase in relative cover of *Corylus cornuta* [beaked hazel]…" (Potter and Moir 1961: 476).

Increasing burning of forests does not necessarily correlate simplistically with agricultural intensification, because thinning forests is also a strategy employed to increase nut yields from oak, hickory, and walnut groves, with the added benefit of increasing hazelnut availability. Thus, the drop-off in hazel use after the Middle

Woodland could indicate "that continuing evolution in subsistence systems ... made horticultural intensification increasingly attractive in contrast to a tree-management strategy" (Asch and Asch 1985a: 354). Whatever the cause of Middle Woodland reliance on hazels, the Mound House assemblage suggests that hazelnut processing was not as frequent there as at other sites in the LIV.

Both thin hickory (*Carya illinoensis*, pecan) and acorns (*Quercus sp.*) are present at Mound House in such small amounts that little can be said about their use. The dearth of acorn is also unusual for the LIV. Acorns, if eaten in large amounts, must be processed to remove tannins. Their absence from the Mound House assemblage may indicate that this processing did not take place at Mound House, as with hickory.

Eastern Agricultural Complex. Of the 945 seeds identified, 57% (570 seeds) were members of the Eastern Agricultural Complex (Figs. 13 and 14). Goosefoot (*Chenopodium berlandieri* and *Chenopodium berlandieri* ssp. *jonesianum*) was the most abundant by far, making up 24.5% of the total identifiable seeds. Of the 263 goosefoot seeds recovered, 192 came from a single sample (94), which shows how much a single rich sample can affect the interpretation of a small assemblage like this. The bases on which domesticated goosefoot (*C. berlandieri* ssp. *jonesianum*) is distinguished from its wild or weedy relatives have been discussed above. To review, criteria applicable to single seeds (as opposed to entire plants) include testas <20 μm, smooth testa texture, and truncate margins opposite the "beak" (Smith and Yamell 2009). The Mound House goosefoot was sorted into two groups based on relative seed coat thickness. Group 1, probable domesticated *C. berlandieri* ssp. *jonesianum*, exhibited testas estimated to be less than 20 μm thick, while Group 2, probable weedy or wild goosefoot, exhibited testas estimated to be greater than 30 μm thick (Fig. 11).

Taking exact measurements of testa thickness requires a scanning electron microscope, but relative thickness is visible at 40× magnification. In addition to relative testa thickness, Group 2 seeds had alveolate (pitted) testas and tended to be much more intact. Group 1 seeds had either alveolate or smooth testas that were often preserved in only small patches on otherwise popped and distorted seeds. The poor preservation of the probable domesticated *C. berlandieri* ssp. *jonesianum* made it impossible to consistently observe margin configuration, but members of Group 2 were universally well preserved and biconvex. These observations allow for a preliminary analysis of the goosefoot populations represented at Mound House (Gayle Fritz, personal communication). In addition, exemplars of each type were examined under a scanning electron microscope (Fig. 10). These representative specimens exhibit seed coat thicknesses of >30 μm and <20 μm, respectively, adding additional support to the low magnification characterizations.

Gremillion (1993) has shown that the relative proportion of weedy and domesticated goosefoot seeds varies considerably between sites in the Eastern Woodlands, with weedy seeds sometimes making up almost half of the assemblage and other times less than 5%. At Mound House, seeds with relatively thin seed coats make up almost two-thirds of the assemblage, but weedy seeds are also well represented.

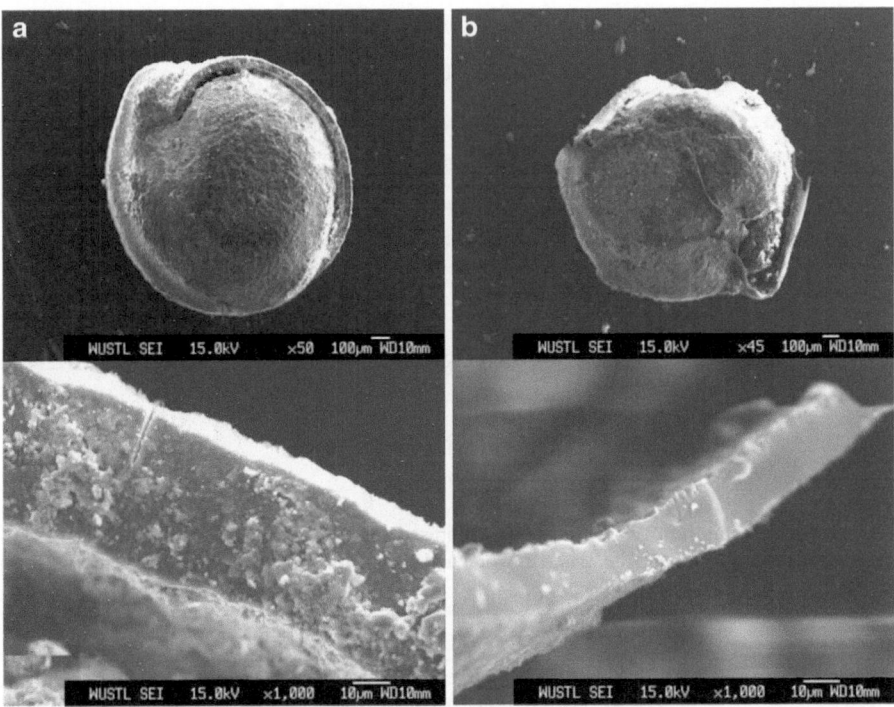

Fig. 10 (**a**) *Top*: Wild/Weedy morph; *Bottom*: seed coat of wild/weedy morph at 1,000×
magnification. (**b**) *Top*: Domesticated morph (*Chenopodium berliandieri* ssp. *jonesianum*); *Bottom*:
seed coat of domesticated morph at 1,000× magnification

Chenopodium berlandieri Morphology at Mound House

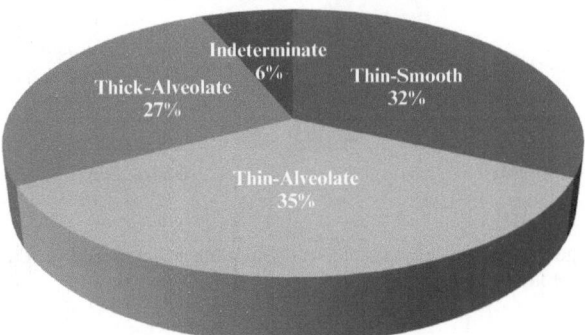

Fig. 11 Proportional representation of different chenopod morphs observed, calculated as a percentage of all chenopods with preserved testa

This pattern suggests the presence of a crop-weed complex where wild *C. berlandieri* continued to colonize clearings where cultivated ssp. *jonesianum* was grown.

Weedy and domesticated seeds occur together in human paleofeces from Salts
and Big Bone Caves, indicating that weedy goosefoot was sometimes harvested and

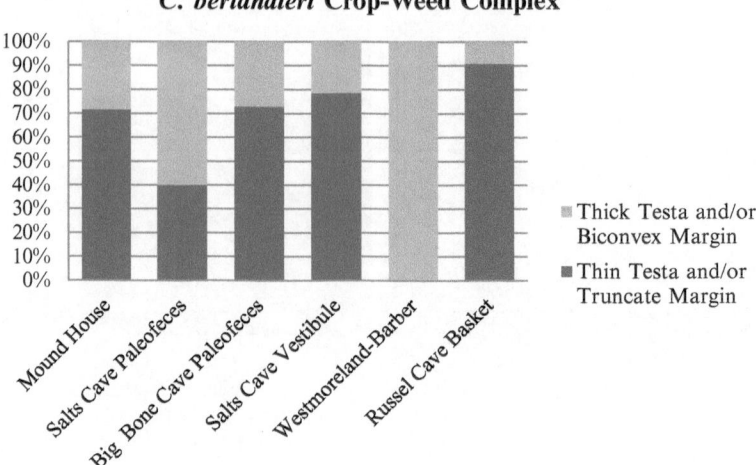

Fig. 12 After Gremillion 1993. Proportional representation of wild/weedy and domesticated chenopods at sites throughout the Eastern Woodlands, in comparison to preliminary results from Mound House

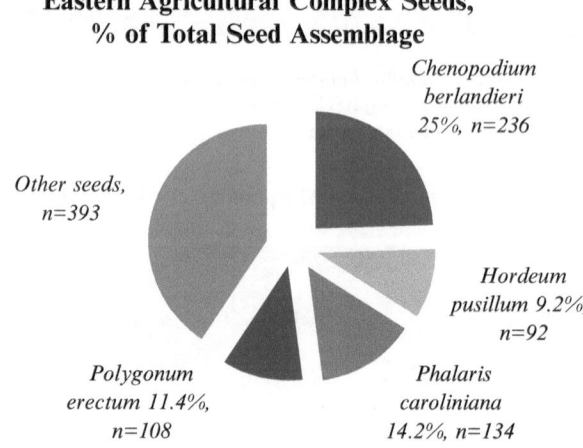

Fig. 13 Percentage of the total identifiable seed assemblage composed on Eastern Agricultural Complex seeds at Mound House

consumed along with crop plants (Gremillion 1993: 499; Fig. 12). In comparison to paleofeces, desiccated collections from storage contexts are more often dominated by domesticated plants—a pattern that provides archaeological corroboration for the seed selection practices that would have led to domestication in the first place (Fritz and Smith 1988; Gremillion 1993: 506). In Fig. 12, the Russell Cave basket assemblage represents this type of context and has a much greater proportional representation of domesticated goosefoot than the other assemblages. As Fig. 12

Ubiquity

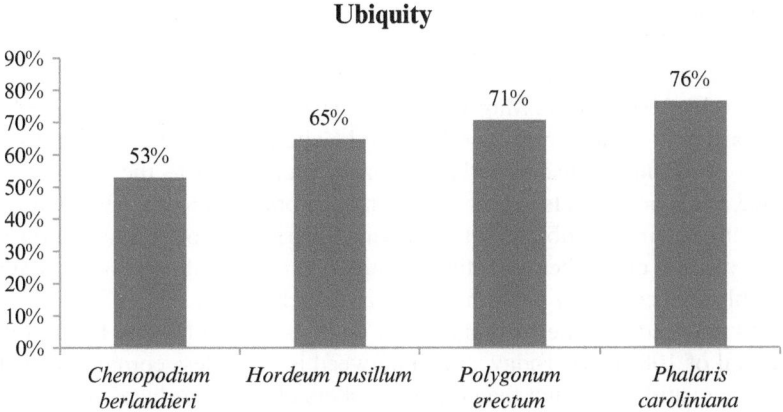

Fig. 14 Ubiquity of Eastern Agricultural complex crops at Mound House

illustrates, the Mound House goosefoot does not have the same overwhelming density of domesticated seeds as Russell Cave assemblage. However, the proportion of domesticated seeds at Mound House may actually be much higher, because only slightly more than half of the goosefoot seeds recovered had observable testa. The other collections in Fig. 12 were all desiccated or derived from paleofeces, rather than carbonized. Unlike the weedy seeds, which are often well preserved when carbonized, the thin testas on carbonized *C. berlandieri* ssp. *jonesianum* tend to be fragmentary, and the seeds themselves are often popped, so unidentifiable perisperm and embryo fragments are more likely to belong to the domesticated type.

It is ecologically logical that goosefoot is the most abundant crop seed at Mound House. The agroecology of the Middle Woodland domesticate is unknown, but wild goosefoot is a very versatile adventive plant. It is absent or rare in poorly drained soils (for example, around receding backwater lakes), but it is still a floodplain weed because it colonizes sand bars and natural levees which are annually scoured by flood waters, as well as upland clearings (Munson 1984: 383). While it is quick to populate the understory of forests as well, it is unlikely that this habitat was preferred by ancient cultivators, because understory goosefoot tends to be tall and spindly, with fewer edible leaves and less abundant seed heads (Smith 1992a: 173). Of the four members of the EAC, goosefoot is the most likely to have thrived if cultivated at or near Mound House.

Experimental gathering of goosefoot seeds in Indiana showed that the peak harvesting season falls between the middle of October and the first week of November. During these weeks, researchers found that by cutting the inflorescences from stands of *C. missouriense* (a species closely related to the domesticate) and then stripping them of seeds, they could gather as much as 2.2 kg of seed per hour (Seeman and Wilson 1984: 305–307).

Maygrass is the next most abundant seed at Mound House, with a total of 134 seeds, or 14.2% of the identifiable seeds. While maygrass is less abundant than goosefoot, it is much more evenly distributed (76.5% ubiquity compared to 52.9%

for goosefoot). Maygrass is less likely to have been cultivated in the vicinity of Mound House. Today, it is usually found in fallow fields and along roadsides in well-drained to dry soils, but is not found along river banks (Fritz 2010). Additionally, maygrass is harvested in late May and early June, when the area surrounding Mound House was likely to have been flooded and the site itself inaccessible on foot. The maygrass at Mound House was thus probably transported to the site after being harvested and processed elsewhere. Accidental carbonization of a few seeds in cook fires better explains the ubiquitous but sparse distribution of seeds than does processing, which at other sites sometimes resulted in masses of carbonized maygrass, presumably burned during parching (Asch and Asch 1985a: 360).

Erect knotweed was the second most ubiquitous seed at Mound House (70.6%) with a total of 108 seeds. Today, erect knotweed is much less common and slower to colonize disturbed ground than other members of genus *Polygonum*, although it does yield more seeds per hour of labor than its close relatives, when it can be located in a stand. It, too, is best harvested in Late October but produces seeds in small amounts all summer (Asch and Asch 1985a: 363, 1985d: 183–187). Erect knotweed is also not a common floodplain plant, although it has been found along roadsides in floodplains. It is characteristically associated with areas disturbed by humans or livestock, seemingly preferring packed earth, including pastures and dirt roads (Asch and Asch 1985d: 186). Of the four EAC crops, it is most difficult to reconstruct the ancient habitat of erect knotweed, because today it is often found as single plants in waste places dominated by a mat of *P. aviculare*, which grows fast and low to the ground (Asch and Asch 1985d: 186). It is more sensitive to mowing than its creeping relatives. Before mowing was introduced, it may have been much more abundant. Erect knotweed may or may not have grown near Mound House, but because it is intolerant of flooding, it would have been less risky to cultivate this plant on the terraces or in the uplands. Also, like maygrass, it is evenly and sparsely distributed throughout the midden, a pattern that does not indicate processing.

The morphology of archaeological knotweed was discussed above. For the Mound House specimens, length and width measurements for complete achenes and observations of texture were recorded. These are presented along with measurements for other Middle Woodland erect knotweed assemblages in the LIV in Fig. 15. Over time, the average size of knotweed achenes (L×W) increased; at Hill Creek, an Late Mississippian site, the mean L×W was 7.4. As Fig. 15 shows, there is relatively little variability within the Middle Woodland assemblages. Seeds as Mound House, Smiling Dan, Crane, and Loy (the latter two are habitation sites on major tributaries of the Illinois River) are all slightly larger than the assemblages from Archie and Massey.

The relative abundance of smooth-pericarp erect knotweed also increases through time; at Hill Creek (a Mississippian site) 100% of the complete achenes are of this type. At Middle Woodland sites, both morphs are represented, but their relative representation between sites is uninformative since the majority of erect knotweed seeds recovered lack complete pericarps. In addition, at some sites (such as Smiling Dan) only a subset of the complete assemblage was classified by shape and pericarp texture in order to ascertain mean lengths and widths for the two morphs. Figure 16

Mean Length and Width (mm)
of Erect Knotweed

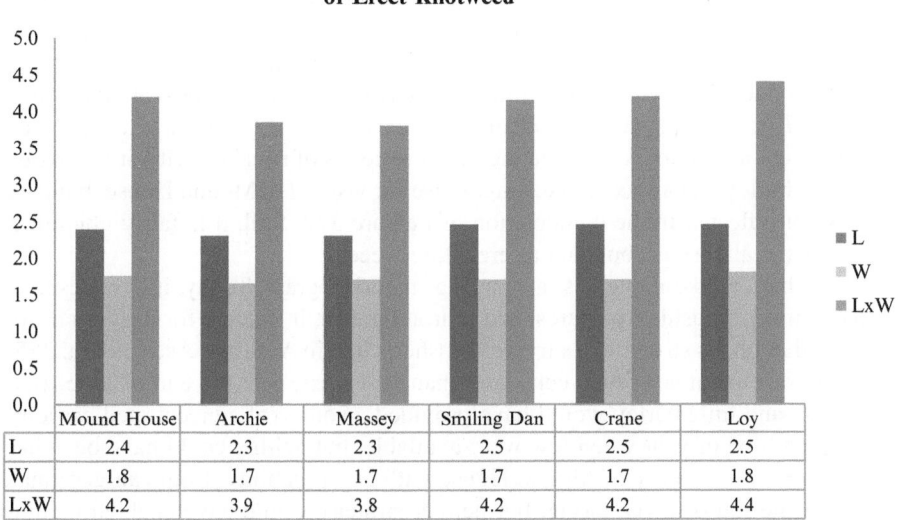

	Mound House	Archie	Massey	Smiling Dan	Crane	Loy
L	2.4	2.3	2.3	2.5	2.5	2.5
W	1.8	1.7	1.7	1.7	1.7	1.8
LxW	4.2	3.9	3.8	4.2	4.2	4.4

Fig. 15 After Asch and Asch 1985b. Comparison of mean erect knotweed (*Polygonum erectum*) achene size between Mound House and other Middle Woodland sites in the LIV

Length X Width (mm)
of Erect Knotweed by Morph

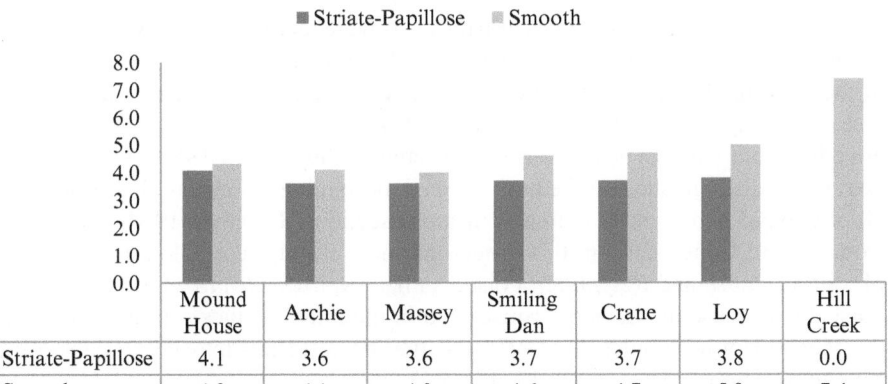

	Mound House	Archie	Massey	Smiling Dan	Crane	Loy	Hill Creek
Striate-Papillose	4.1	3.6	3.6	3.7	3.7	3.8	0.0
Smooth	4.3	4.1	4.0	4.6	4.7	5.0	7.4

Fig. 16 After Asch and Asch 1985b. Comparison of mean erect knotweed (*Polygonum erectum*) size between two morphological varieties, including Mound House and other Middle Woodland sites in the LIV. The Hill Creek assemblage dates to the Mississippian period and possibly represents a morphologically domesticated variety of erect knotweed; it is included for comparison

compares these measurements. The Mound House assemblage stands out as somewhat unusual because its striate-papillose morph is larger on average than that from any of the other assemblage. Here, Hill Creek is included for reference to what may be a domesticated form a millennium later.

Little barley makes up 9.6% of the seed assemblage with a total of 92 seeds. It should be noted that of the 292 seeds identified by Marjorie Schroeder from the excavations around Mound 1, 92 were little barley, doubling the total from the site as a whole (Schroeder 1998: 1986–1987). Little barley was the only seed recovered in any notable quantity from the Mound 1 excavations, the next most abundant type being *Polygonum* sp. with a total of six. Because the deposits in and under the mound itself are more likely to be the direct residues of ritual activity, it is possible that little barley had a special meaning or use for visitors to Mound House, but if so, this is not reflected in the composition of Feature 379. Still, it is fairly ubiquitous (64.7%) and almost as abundant as erect knotweed.

Little barley, like maygrass, is a spring-maturing grass. Today, it is widespread and colonizes roadsides, pastures, and railroad tracks. It is noted for thriving in soil where other plants struggle to survive (Mosher, cited in Asch and Asch 1985a: 385). In Illinois, it matures a few weeks later than maygrass, near the end of June. Both maygrass and little barley would have provided a source of concentrated carbohydrates at a time of year when few were available, but neither could have been harvested from the vicinity of Mound House with any reliability. Both can germinate in either the fall or early spring, but neither strategy would have enabled them to survive spring floods. However, if visits to Mound House occurred in midsummer, as the current interpretation of ritual activity suggests, then little barley and maygrass, recently harvested and abundant, would have been a logical food to bring to the site. More speculatively, as spring resources available at the same time as the height of flooding, they may have been meaningful within the context of world-renewal ceremonialism (Fritz 2010).

Other edible plants. Several types of fruit seeds were recovered from Mound House. Sumac (*Rhus* sp.) was the most abundant of these, with a total of 53 seeds or 5.5% of the identifiable seeds. Several closely related species of sumac are native to the area, including *R. glabra*, *R. copallina*, *R. typhina*, and *R. aromatica*, all of which are edible, but with berries of varying sweetness (Steyermark 1981). These cannot be reliably distinguished on the basis of seed morphology alone, as all are similarly elliptic ovoid and slightly flattened in cross section (Borojevic 1994). The seeds from Mound House fall into this group and most closely resemble reference samples of *R. glabra* and *R. typhina*. Poison sumac (*R. toxicodendron*) is easily distinguishable because its seeds are constricted in the middle—these are not present in the Mound House samples. Depending on the species, fruits may be available from May to November (Steyermark 1981: 1002).

Historically, sumac berries were soaked in water to prepare a sour beverage. There are also records of sumac being stored for the winter, so they are not a good indicator of seasonality. Sumac fruits were also widely used for medicine, especially to stop bleeding and to treat urinary tract infections. The leaves were smoked with tobacco, and the roots were used to produce a yellow dye by the historic Omaha and Winnebago (Gilmore 1977: 48; Steyermark 1981: 1000). Entries for the medicinal uses of sumac consume almost four pages of Moerman's encyclopedic *Native American Medicinal Plants: An Ethnobotanical Dictionary* (2009: 410–413).

Its unusual abundance at this ritual site may be due to its many medicinal, ceremonial, and symbolic uses.

Wild grapes (*Vitis* spp.) were also common at Mound House. A total of 35 were recovered with a ubiquity of 41%. These show considerable variation and probably represent more than one of the seven wild species known to grow in the area. Three of the seeds bore a close resemblance to reference samples of *V. cinerea*. *V. cinerea* (Grayback grape) "occurs in low woods and alluvial soils along streams" and produces small (4–7 mm), sweet fruits during September and October (Steyermark 1981: 1037). These same characteristics are common to several species, although some fruit as early as July (*V. aestivalis*) and others produce fruits as large as 12–25 mm (*V. rotundifolia*). The latter are the ancestors of the cultivar Muscadine grapes used to make Missouri wines. They are known to have been collected by various Southeastern people, dried and stored or used to produce a sweet fruit juice (Gilmore 1977: 50; Steyermark 1981: 1035–1041).

A minimum of three American lotus (*Nelumbo lutea*) seeds were recovered. This species would have grown in backwater lakes in the immediate vicinity of Mound House and is the only wetland species identified. It flowers between late June and September, with nut-like seeds becoming available in late summer. Early in the season, these can be collected and eaten raw, while kernels removed from ripe seeds in the autumn were either cooked or ground in flour. Tubers also become available in autumn; none of these were recovered (Steyermark 1981: 668). Both the seeds and tubers are known to have been valued food sources by historic American Indians and continued to be coveted by settlers and trappers through the seventeenth century, when one European admirer described the seeds as "just like chestnuts" (quoted in Asch and Asch 1985a: 387). Gilmore (1977: 27) reported that American Indians along the Missouri River considered American lotus plants to be "invested with mystic powers" and recounted tales which attribute human qualities to their banana-shaped tubers, but they are not reported to have any medical use.

Five pokeweed (*Phytolacca americana*) seeds were identified. Pokeweed is still collected as a spring potherb, but its red berries are widely considered poisonous, despite pharmacological evidence to the contrary (Asch and Asch 1985a: 385). Pokeweed flowers from May to October and can occur in almost any kind of soil or physiographic location (Steyermark 1981: 630). Sauer's (1950) pokeweed overview reveals that the berries were once used to make tarts and pies by Euro-Americans and that, once introduced to Europe, the berries were used to color wine and spirits. Similarly, various historic American Indian groups used the berries to make water-soluble dye. The berries were also considered a treatment for arthritis by the Cherokee and rheumatism by the Rappahannock and Euro-Americans, while the roots were used for a variety of ailments across the eastern half of the continent (Moerman 2009: 353; Steyermark 1981: 630). Whether as food or medicine, we know that the ancient inhabitants of the Eastern Woodlands consumed pokeweed berries, because they were found in the paleofeces from Salts Cave (Yarnell 1969: 44).

A minimum of 11 persimmon (*Diospyros virginiana*) seeds were recovered from 17% of the samples. This tree bears fruit in the early autumn which can be eaten fresh or dried and stored. One seed each of *Rubus* sp. and *Vaccinium* sp. were also identified. *Rubus* includes blackberries, raspberries, and dewberries, which are all so closely related that it can be difficult to distinguish even between live plants. Generally, all species fruit in late summer and thrive along streams and forests edges, and all produce sweet, edible fruits (Steyermark 1981: 834). *Vaccinium* includes blueberries, deerberries, and gooseberries, all of which prefer acidic, rocky soil, and full sun and many of which produce sweet berries during the summer (Steyermark 1981: 1160–1164).

Other seeds. It is common for paleoethnobotanical reports to include a heading "Noneconomic seeds," but in this case such a heading seems unwarranted since some of the remaining taxa are more abundant than species with established uses. If abundance can be used to argue for cultivation, it is at least sufficient to argue for usefulness. Purslane (*Portulaca oleracea*) is one example. A total of 59 purslane seeds were recovered, and with a ubiquity of 64%. This taxon was as widely distributed as some members of the EAC, an unusual abundance in comparison to the contemporaneous LIV sites Smiling Dan, Archie, and Massie, where it was entirely absent (Asch and Asch 1985a: 357–358; Asch and Asch 1985c: 181), and other Middle Woodland sites in the American Bottom where it also has not been reported (Simon and Parker 2006). Various species of purslane are edible and colonize open spaces, so it is possible that purslane was collected and eaten at Mound House. Similarly, biconvex *Polygonum* sp. seeds (probably smartweeds) were not uncommon ($n = 17$, ubiquity 65%). Many members of the genus are weedy plants with edible seeds that are very common in bottomlands (Murray and Sheehan 1984). Twenty-five seeds of the nightshade genus (*Solanum* sp.) with a ubiquity of 41% were identified. This genus contains poisonous, medicinal, and edible plant species, and it is not clear which of these uses the Mound House *Solanum* were put to, if any. Black nightshade (*Solanum americanum*) is the most widespread species, ranging from Maine to Texas. It produces edible black berries and often grows along streams, roads, and other disturbed places (Steyermark 1981: 1312).

Small panicoid grass seeds are also abnormally abundant at Mound House. I have provisionally labeled these "Panicoid type" because they lack features that allow identification beyond tribe. A total of 126 were recovered with a ubiquity of 52%, making Panicoid type the most abundant seed apart from goosefoot and maygrass. The amount of non-EAC grass seeds at Mound House is anomalous in comparison to every other assemblage in the valley. The area around Mound House was recorded as a floodplain prairie at the time of European settlement, and several panicoid species are components of this now rare ecosystem. The Mound House Panicoid type could simply reflect the surrounding vegetation, especially if it was used to create temporary shelters or to thatch structures that were subsequently burned or if its seeds were harvested for food here but not elsewhere (Asch and Asch 1985a: 390).

It is also possible that the Panicoid type at Mound House had a ritual use. Geoarchaeological investigations of Mound 1 at Mound House have revealed that the mound was constructed using stacked, inverted sod blocks. The cultivation or harvesting of sod needed as a construction material may account for the preponderance of grass seeds at Mound House, but further investigations of the sod used in construction are necessarily to support this speculation. We still do not know what plant or plants formed these sods or if they were cultivated or merely harvested from wild prairies (Van Nest 2006).

Several weedy taxa were present in small amounts, including one seed each of barn grass (*Echinochloa* sp.) and verbena (*Verbena* sp.), two of pigweed (*Amaranthus* sp.), and a fescue-type grass (Poaceae subfamily Festucoid), plus nine other seeds identifiable only as grasses (Poaceae). The remainder of the seeds ($n=27$) were unidentifiable. In addition, 348 unidentifiable seed fragments were also pulled from the samples, but these are not included in any calculations of seed density or ratios of seed frequency.

Comparative Analysis

The patterning of botanical material between sites can help clarify our understanding of how Mound House fit into the ecological and cultural landscape of the LIV. Given that plant-related activity varied between sites in the valley, and variability can be seen in the patterning of botanical refuse, this variability can help us understand the trend of agricultural intensification during the Middle Woodland period. Mound House, Smiling Dan, and the Massey phase sites are representatives of habitation in the three major physiographic regions of the valley, the floodplain, bluff base, and upland creeks. Botanical assemblages from Peisker and Napoleon Hollow are the residues of formal visitations to floodplain and bluff-base mound complexes, respectively. A comparison between these two sites and Mound House may show how (or if) subsistence-related activity at mound complexes varied according to their ritual function.[1]

Non-mound Habitation Sites

This comparison between Mound House and non-mound habitation sites in the LIV (Smiling Dan and Massey phase sites) proceeds from a set of expectations about how Mound House refuse may differ from that at other habitation sites. If people

---

[1] A note on comparability: It is unclear from the published descriptions of methods (Asch and Asch 1985a, 1985c, 1986; Calentine 2005; Staab 1984) whether or not any attempt was made to develop Seed Number Estimates for taxa which are identifiable as fragments. However, it seems likely that for *Chenopodium berlandieri*, at least, some extrapolation from raw counts was employed. These seeds often split into three parts (two halves of the testa and the perisperm/embryo), which are all diagnostic, making a raw count of fragments extremely misleading as an indicator of total number of seeds. The only other taxon discussed comparatively for which SNEs differ from total counts in my analysis is sumac (*Rhus sp.*). Since sumac is relatively abundant at Mound House, I have used SNEs in the comparison order to err on the side of caution and not overstate the variability in sumac frequency between Mound House and other sites. Total number of seeds for each assemblage includes unidentifiable seeds, but not seed fragments, since these would only compound the problem of calculating minimum numbers of seeds uniformly between sites.

visited Mound House seasonally to participate in a larger symbolic community, they would have engaged in different activities at Mound House than at their home-steads. If seeds and nuts were not harvested, processed, or stored at Mound House, then people would have needed to bring stored seeds to the site. They may also have eaten differently at Mound House than at home or have consumed more medicinally or ritually useful plants. This should be reflected in the garbage they created. The following set of hypotheses is meant to test whether or not people from different residential communities brought food to Mound House, facilitating an exchange of seed stock and agricultural knowledge:

(a) Null Hypothesis: Plant-related activities at Mound House were similar to those at Middle Woodland non-mound habitation sites in the LIV. Seed crops and nuts were harvested, processed, and stored at Mound House.

Expectations:

- There will be no difference in the percent frequency or distribution of crop seeds between Mound House and other habitation sites.
- There will be no difference in the diversity and type of species represented at Mound House.
- Nutshell will be equally abundant at Mound House.

(b) Alternative Hypothesis: Mound House was the site of periodic intercommunity integration. Nuts and seeds *were not* harvested, processed, or stored at Mound House. Time spent at Mound House was partly dedicated to sharing meals and exchanging knowledge. Food residues present at Mound House represent trans-ported stores harvested and processed at habitations elsewhere.

Expectations:

- Percent frequency of crop seeds will be lower at Mound House. There will be less variability in the distribution of seeds because the range of plant-related activity was limited to cooking. Masses of seeds will be absent, as these are the result of processing accidents or of storage
- Species that are uncommon or absent at habitations may be present or abun-dant at Mound House due to special ritual, technological, or social uses. Conversely, species that are common at habitation sites may be absent at Mound House because they are a result of agricultural activity that did not take place at Mound House.
- Nutshell will be less abundant overall because nuts were processed elsewhere.

Smiling Dan

Smiling Dan is comprised of three houses clustered on the alluvial fan of a tributary stream where it enters the main valley. The Smiling Dan samples were taken from three large middens and many smaller pits. The botanical assemblage from Smiling

Ubiquity

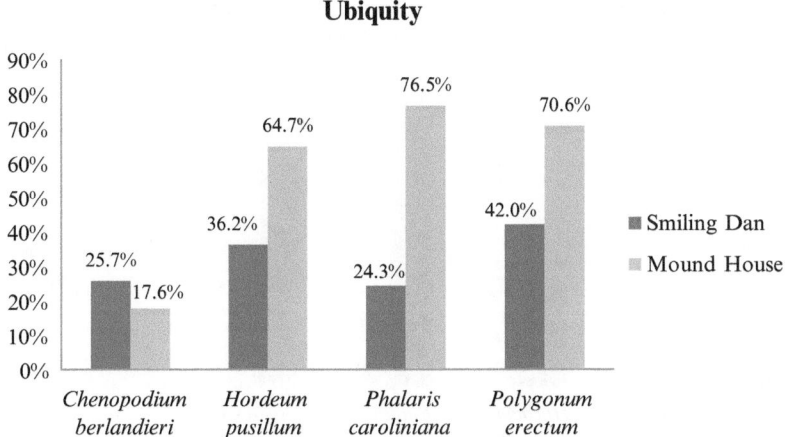

Fig. 17 Ubiquity of EAC crop seeds at Mound House and Smiling Dan (Asch and Asch 1985a)

Dan is much larger than that from Mound House. A total of 12,000 l of sediment were sampled, as opposed to 139.5 l from Mound House. Despite the greater number of sampled features at Smiling Dan, the difference between assemblages is primarily one of volume. Some of the pits at Smiling Dan were presumably used for storage at one point but were eventually filled with midden deposits. According to Asch and Asch, "Statistics based on median values show that features typically resembled the midden rather closely" (1985a: 398). While on average the pits had a greater seed density, these deposits are fundamentally midden. There is no evidence from any of the pits for *in situ* burning of stored seeds.

Expectation A: *Relative Abundance and Distribution of Crop Seeds*. I expected that if stored seeds were brought to Mound House, rather than prepared at the site, seeds would be less abundant at Mound House than at Smiling Dan. Processing seeds usually involves parching them to remove chaff. For little barley, parching has been shown to greatly reduce the time and labor required to remove the inedible plant parts from the grain (Gasser 1982: 220–221). Parching can lead to the carbonization of large numbers of seeds that are subsequently discarded. For example, at Smiling Dan, 64% of the maygrass seeds recovered came from one context, Feature 61. The maygrass in Feature 61 was scattered throughout the midden in "small clumps" of adhering grains (Asch and Asch 1985a: 362). The most likely explanation for these concentrations is a parching accident. If seeds were not harvested and processed at Mound House, inputs from this source would be entirely removed. Smaller inputs from all stages of plant procurement, from harvesting, processing, and storing, would also be removed. Only accidents occurring during cooking would add carbonized seeds to the midden deposits at Mound House.

Figure 17 shows ubiquity of EAC crops at Mound House and at Smiling Dan. This distribution supports the expectation that seeds will be more evenly distributed (i.e., more ubiquitous) at Mound House. Seeds and crop seeds, in particular, are

**Eastern Agricultural Complex Seeds, as % of
Total Seed Assemblage**

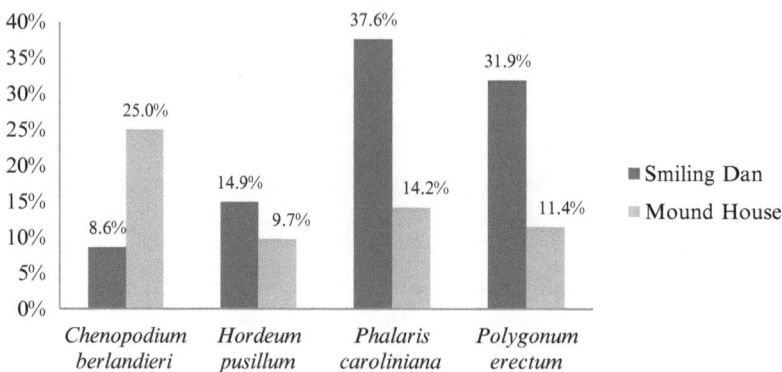

Fig. 18 EAC crop seeds as proportion of the total seed assemblage (all identified seeds plus unidentifiable seeds, but not seed fragments) (Asch and Asch 1985a)

found throughout the Feature 379 midden. At Mound House, seeds are present in 100% of the samples, while at Smiling Dan they are only present in 76%. This may reflect a greater range of activity at Smiling Dan than at Mound House; some midden deposits contain no seeds because they were not near crop processing or cooking areas. The absence of seeds from almost a quarter of the Smiling Dan samples makes it difficult to use site-wide standardized density as a standard for comparison, because thousands of liters of sediment from seed-free contexts dilute the densities for Smiling Dan.

Using a ratio of EAC seeds to all seeds recovered as a basis for comparison eliminates bias introduced to calculations of density by the sheer volume of seed-free samples processed at Smiling Dan. Crop seeds make up a much larger proportion of the total seed assemblage at Smiling Dan (Fig. 18). This strongly supports the alternative hypothesis that seeds were transported to Mound House rather than harvested and processed there. If all edible seeds are more or less equally likely to be burnt in the process of cooking, seed crops that are harvested for grain are more likely to be better represented at sites where they were parched, threshed, and stored than at a site where they were merely consumed.

The only crop seed that made up a greater part of the assemblage at Mound House than at Smiling Dan was goosefoot. This may be partially the result of the small sample size at Mound House, which gives greater weight to extremely rich samples (such as sample 94 with its 192 *Chenopodium berlandieri* seeds). But since goosefoot is the only member of the EAC that might have been grown in frequently flooded habitants, it is also possible that harvesting and processing of goosefoot was carried out at Mound House. Although exact proportion of wild and domesticated varieties are not published, the Smiling Dan assemblage is described as nearly evenly split between the two types (Asch and Asch 1985a: 372), meaning that domesticated goosefoot is substantially better represented at Mound House than at Smiling Dan.

Although this result is preliminary due to the lack of exact seed coat measurements for the Mound House goosefoot, it tentatively supports the notion that seeds recovered from Mound House were transported stores, that is, those seeds that had been selected by humans either intentionally or as a result of their domestication syndrome.

Rarefaction and Significant Differences in Abundance: Testing Expectation A. The Smiling Dan assemblage is the largest botanical assemblage ever analyzed from the LIV. The entire site was floated, including house floors, middens, hearth, and pits. Because this collection is so large and comprehensive, in this analysis, I will treat it as a complete representation of the LIV's Middle Woodland household agroecology. The Mound House assemblage is much smaller. Only 139.5 l of sediments were floated, and all of these come from a single midden. My exploration of the data suggested that there were differences in the relative abundance of plant taxa between the two sites, but it is possible that the apparent differences are a result of sampling error. In order to ascertain whether or not crop seeds are actually less abundant at Mound House than at Smiling Dan, I have borrowed and modified a statistical method developed by community ecologists.

Rarefaction is a method developed by ecologists (Simberloff 1972) to evaluate whether or not a given sample size is sufficient to reflect the true diversity of the ecosystem being sampled. Usually, rarefaction curves are used to derive an expected number of species for a given sample size (in ecology, usually area or distance traversed). This is accomplished by resampling from the distribution of a larger initial dataset from the same type of ecosystem. A smaller subsample of individuals is drawn from the large dataset in proportion to its representation therein. Species that are rare in the large dataset disappear from the rarefied datasets as the subsample size decreases. This yields an expected number of species for a given coverage area and allows ecologists to gauge how large of an area they need to survey in order to capture the full diversity of the study area, estimate at what point further sampling becomes redundant, and make meaningful comparisons of species diversity between a large study area and a small study area. This method has also been used by ethnobotanists, who use number of informants in place of area to evaluate the adequacy of their sampling efforts (Begossi 1995).

The rarefied datasets used to generate rarefaction curves can also be used to test my hypothesis. Using the R Programming Environment, I generated 1,000 rarefied datasets of the same size as the Mound House seed assemblage ($n=924$) drawn from the Smiling Dan seed assemblage. I then counted the number of times that the rarefied population for each taxon was greater than or equal to the Mound House population and the number of times that it was less than the Mound House population. These counts yielded what I refer to as greater/equal abundance and lesser abundance indices. These indices are values from 0 to 1 that reflect the percentage of instances (out of 1,000 datasets) in which a taxon in the Mound House assemblage is more or less abundant than the same taxon in the rarefied subsamples. I considered values of greater than 0.95 to be significant.

The distribution of lesser abundance index values is shown in Fig. 19. There are many very low values, which reflected the even but low density distribution of a variety of fruit and weed seeds at Mound House: these taxa are not consistently less

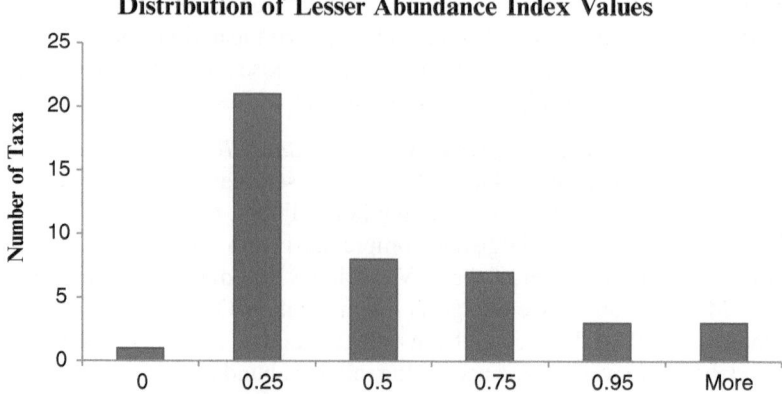

Fig. 19 A histogram of Lesser Abundance Index values. This shows how many taxa are less abundant at Mound House than in 1,000 rarefied samples drawn from Smiling Dan. Many taxa (21 out of 43) were only less abundant in 0–25% of rarefied samples. The three taxa that were more abundant in more than 95% of rarefied samples than at Mound House were *Hordeum pusillum*, *Phalaris caroliniana* and *Polygonum erectum*

well represented at Mound House than in random subsamples of the Smiling Dan dataset. Out of the 43 taxa identified at one or both of the sites, the only taxa with a lesser abundance index of greater than 0.95 are *Hordeum pusillum* (0.962), *Phalaris caroliniana* (1.00), and *Polygonum erectum* (1.00), three out of four of the crop plants present at Mound House. The fourth, *Chenopodium berlandieri*, has a lesser abundance index of 0.884—high but not significant. In other words, if it was possible to randomly draw a seed assemblage the size of the Mound House assemblage from Smiling Dan, greater than 950 times out of 1,000, it would contain a greater concentration of crop seeds. Using this modified version of rarefaction analysis, it is possible to show significant differences in abundance between two assemblages of very different sizes.

Expectation B: *Diversity*. Figure 20 shows the percentage of seeds made up by several taxa that were known ethnographically to have ritual or medicinal uses among Eastern or Plains American Indians. American lotus (*Nelumbo lutea*) was present in very small amounts at both sites ($n=3$ at Mound House, $n=8$ at Smiling Dan), and its slightly greater relative abundance at Mound House might be explained by the proximity of the site to wetland habitats. *Rhus* sp., *Solanum* sp., and *Vitis* sp. all contain species that are both edible and known to have been used medicinally by historic eastern Native American peoples and all are relatively much more abundant at Mound House than at Smiling Dan. A mid- to late summer period of occupation is another possible explanation for the unusual abundance of summer fruit seeds at Mound House. The fruits could have been gathered and consumed fresh in the vicinity of Mound House.

The Smiling Dan assemblage is much more diverse than the Mound House assemblage, and some of this diversity may be the result of the enormous sample

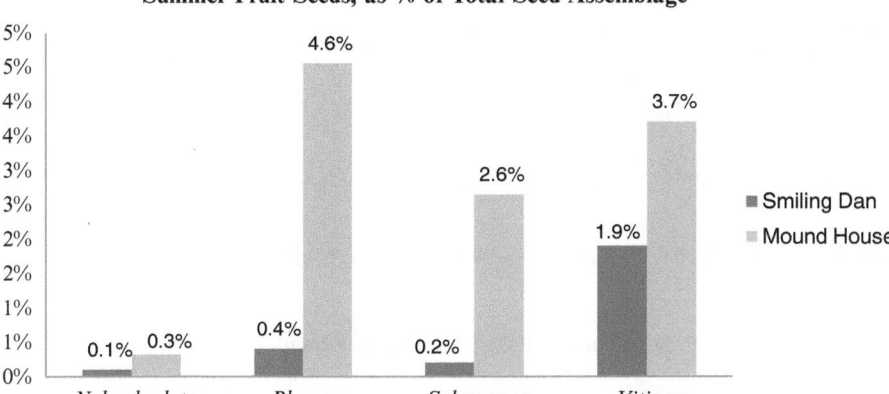

Fig. 20 Summer fruit seeds as proportion of the total seed assemblage (all identified seeds plus unidentifiables, but not seed fragments) (Asch and Asch 1985a). American lotus (*Nelumbo lutea*), sumac (*Rhus* sp.), and nightshade (*Solanum* sp.) have ethnographically recorded medicinal or ritual uses among eastern American Indian people; wild grape (Vitis sp.) does not

size at Smiling Dan. The Smiling Dan assemblage includes cultivated plants that are common at hamlets and absent so far at Mound House. These include squash (both *Cucurbita* sp. and *Lagenaria siceraria*), sunflower (*Helianthus annuus*), marshelder (*Iva annua*), and members of the bean family (Fabaceae). It may be that these plants are yet to be discovered at Mound House, but their absence also supports the alternative hypothesis since less visible crop plants would be less likely to appear at a site where agricultural production did not take place.

The difference in abundance of grass seeds between the two sites is also striking. Categories of identification vary between the two sites, so I have combined all members of family Poaceae as a basis for comparison: at Smiling Dan grass seeds make up only 0.5% of the total seed assemblage, while at Mound House they make up 14.4%. Possible ecological, technological, and ritual explanations for the concentration of grass seeds at Mound House were reviewed above.

Expectation C: Nutshell. My third expectation was that nutshell would be less abundant at Mound House than at Smiling Dan. This expectation was not borne out: in fact, nutshell density at Mound House is about three times higher than at Smiling Dan (standardized density at Mound House=0.18 g of nutshell per liter and only 0.06 g per liter at Smiling Dan). As with seed density, this disparity is partly a reflection of differing sample size. However, ubiquity was more comparable between the two sites for nutshell than for seeds (Fig. 21). Another common measure of relative nutshell abundance is the ratio of nutshell-to-wood charcoal. By this measure, too, nutshell is very abundant in Mound House garbage (Mound House=1.3 g of nutshell per gram of wood, Smiling Dan 0.78 g of nutshell per gram of wood). Evidently, nut processing was taking place at Mound House.

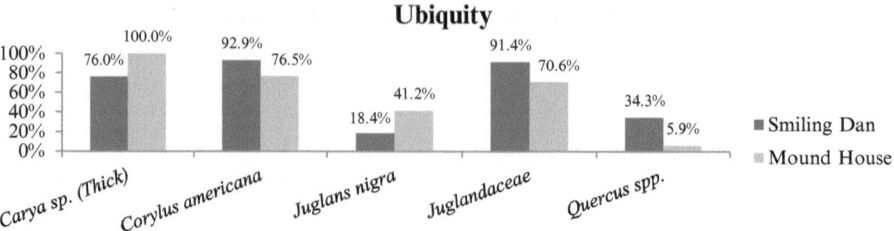

Fig. 21 Ubiquity of nutshell fragments >2 mm (Asch and Asch 1985a)

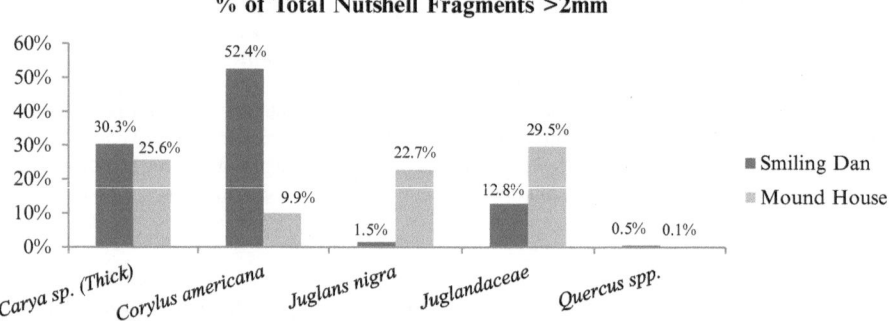

Fig. 22 Nutshell taxa as a proportion of the total nutshell assemblage, all fragments >2 mm (Asch and Asch 1985a)

The relative abundance of nutshell between the two sites may indicate ecological or behavioral difference (Fig. 22). The most striking difference between the two assemblages is that hazelnut makes up more than 50% of the Smiling Dan assemblage and less than 10% at Mound House. Hazels are associated with burning and may have been harvested from fallow fields. Their striking underrepresentation at Mound House is another indication that the harvesting of agricultural plots was not a part of activity at the site, but they are also unlikely to have grown in the immediate vicinity of the Mound House. Black walnut shell is rare at Smiling Dan, while at Mound House it makes up almost 30% of the assemblage. Ecological and technological explanations for the abundance of black walnuts at Mound House were discussed above. While both acorns and hickory nuts need to be laboriously processed to make edible foods, black walnut meats can easily be eaten straight out of the shell. Like the summer fruits, walnuts might have been an expedient food to process and consume during visits to Mound House, but only if these occurred in the autumn.

Massey Phase Sites

Expectation A: *Relative Abundance and Distribution of Crop Seeds*. The assemblages from Archie and Massey are combined for convenience. The two assemblages

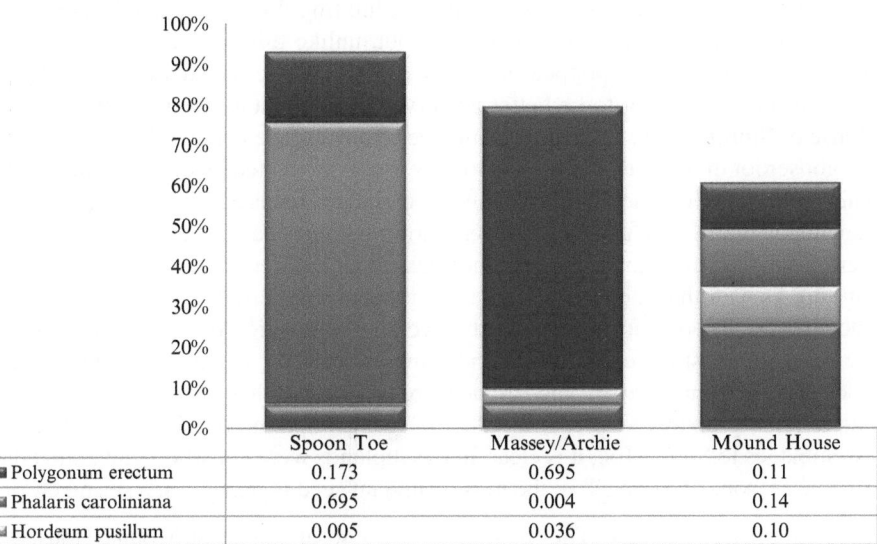

Eastern Agricultural Crop Seeds as % of Total Seed Assemblage

	Spoon Toe	Massey/Archie	Mound House
▪ Polygonum erectum	0.173	0.695	0.11
▪ Phalaris caroliniana	0.695	0.004	0.14
▪ Hordeum pusillum	0.005	0.036	0.10
▪ Chenopodium berlandieri	0.054	0.057	0.25

Fig. 23 EAC crop seeds as proportion of the total seed assemblage (all identified seeds plus unidentifiables, but not seed fragments) (Asch and Asch 1985c; Calentine 2005)

come from two homesteads 750 m apart on the low bluffs overlooking Sandy Creek, about 26 km to the east of the Illinois River. All features and middens were sampled, for a total of 72 samples or 1,600 l of sediment floated. Just as at Smiling Dan, this sampling strategy resulted in many seed-free samples: ubiquity for all identifiable seeds was only 61% as opposed to 100% for Feature 379 at Mound House (Asch and Asch 1985c). Spoon Toe is a similar site about 5 km to the south of Massey and Archie, where 129 samples or a total of 1,233 l of sediment were sampled (Calentine 2005).

Crop seeds made up 79% of all identifiable seeds at Archie and Massey and 93% at Spoon Toe, as opposed to 77% at Smiling Dan and only 59% at Mound House. At both upland sites, one seed type makes up the majority of the assemblage (Fig. 23). At Archie and Massey, erect knotweed (*Polygonum erectum*) was by far the most abundant seed. This one taxon alone made up 70% of the assemblage. Also like Smiling Dan, one particular context accounts for much of this abundance. Seeds from a single clay-sealed layer of one of the Massey pits made up 80% of the entire assemblage, and 73% of these were knotweed achenes (Asch and Asch 1985c: 197). This comes to about 1,000 seeds in a single 10 cm by 0.5 m² area. This lens was unusually well preserved because a layer of clay was packed above it and then used as a hearth, carbonizing and sealing the garbage underneath. The concentration of knotweed in this context, like the clusters of maygrass at Smiling Dan, is suggestive of processing.

Most of the goosefoot from the two sites was also recovered from this context. Asch and Asch (1985c: 185) describe it as exhibiting "seed coat remnants that are smooth, thin, dull and sometimes crazed," but, unlike thin-coated types from elsewhere in the site, rarely popped or distorted. Perhaps the unusual context allowed domesticated goosefoot to be better preserved in this feature than at either Mound House or Smiling Dan. Like the assemblages from storage contexts in rock shelters, the goosefoot in the sealed pit seems to have been dominated by domesticated seeds. This preliminary characterization of the assemblages lends additional support to the conclusion that Archie and Massey were farming homesteads where cultivated seeds were processed and stored. Archie and Massey also differ from Mound House and Smiling Dan in that the assemblage is dominated by the fall-maturing crops erect knotweed and goosefoot. However, the overall composition of the assemblage and the presence of storage pits, hoes, and a unique ceramic assemblage suggested to Asch and Asch that Massey and Archie were inhabited at least from late spring until late fall and probably year round (1985c: 210). Comparing the Massey and Archie assemblages to Spoon Toe, it is clear that the uplands were not strictly inhabited during one season. At Spoon Toe, the most abundant seed is maygrass, a spring crop.

Expectation B: Diversity. The medicinal plants so abundant at Mound House are either absent or extremely rare at all three upland sites. Sumac is present at Archie and Massey, but makes up only 0.8% of the seed assemblage, as opposed to 4.6% at Mound House. In other ways, however, the upland sites are more diverse and more similar to Smiling Dan. Oily seeds are present at all three sites, including squash, marshelder, and sunflower (Asch and Asch 1985c: 185–187; Calentine 2005: 164). There is also a greater diversity of weed seeds with no obvious use, such as bedstraw, tick trefoil, poison ivy, and thistle (Asch and Asch 1985c: 191). As at Smiling Dan, this may reflect a wider variety of agricultural activity than at Mound House.

Expectation 3: Nutshell. The overall density of nutshell at Archie and Massey is higher than at Mound House, but Spoon Toe nutshell density was lower than either (Archie and Massey: 0.47 g per liter, Spoon Toe: 0.03 g per liter, as opposed to 0.18 g per liter at Mound House). This may not be the best measure of the intensity of nut processing, but it does put Mound House within the range of variability at other habitation sites and indicates that nut processing was likely carried out there.

The relative abundance of nutshell by taxa between the three sites is more informative (Fig. 24). At Spoon Toe, nutshell was less abundant overall, and more than 60% was made up of thick hickory. At Archie and Massey, more than 50% of the assemblage was made up of hazelnut. These distributions might be affected by ecology or agricultural activity. At Spoon Toe, the majority of the wood recovered was also hickory, suggesting that the site was located in a grove of hickory trees providing convenient dead wood and nuts (Calentine 2005: 67). At Archie and Massey, 97% of the identified wood was also hickory, but historically the sites were located at the edge of the prairie in open canopy barrens, the ideal environment for hazel (Asch and Asch 1985c: 163–176). Alternatively, hazelnuts may have been harvested from fallow fields cleared by fire, in which case their relative abundance at all three sites examined so far in comparison to Mound House supports the hypothesis that agricultural production did not take place at Mound House.

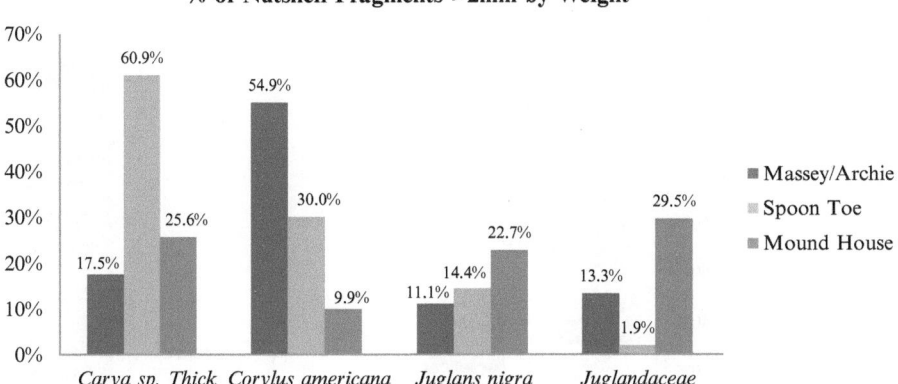

Fig. 24 Nutshell taxa as a proportion of the total nutshell assemblage, all fragments >2 mm (Asch and Asch 1985c; Calentine 2005)

Other Mound Centers

Current models posit that there is a difference in site use between floodplain mound centers and bluff top mound centers, with the former serving as locations for multi-community gatherings and the latter being primarily used by a single cluster of habitations for the burying of their dead (Buikstra et al. 1998; Ruby et al. 2006). Extensive excavations at Napoleon Hollow, a habitation area associated with the bluff-top Elizabeth mound group, revealed very different refuse patterning than at habitations sites like Smiling Dan and led excavators to hypothesize that those who lived at or visited Napoleon Hollow were being provisioned by others (McGimsey and Wiant 1986: 540–541). More limited sub-mound excavations at Peisker led Staab (1984: 239) to conclude that the site was occupied seasonally, most likely in the autumn, and that "the food quest was a priority activity." Both these conclusions and the distinction that has been drawn between flood plain and bluff-top mound centers can be reexamined in light of the Mound House data.

Peisker

Excavations at Peisker were more extensive than those at Mound House: they removed the entire largest mound (Mound 3), whereas Mound 1 at Mound House has been left largely intact. The sub-mound features sampled at Peisker were occupied between 40 BCE and 250 CE (Staab 1984: 44); they are contemporary with, or perhaps slightly later that, the sub-mound activities at Mound House (90 BCE– 40 CE ±70). Also like Mound House, the traces of iterative circular structures underneath Mound 3 indicate several episodes of occupation and construction at the site before the mound was built. Hearths and pits are scattered within and around these

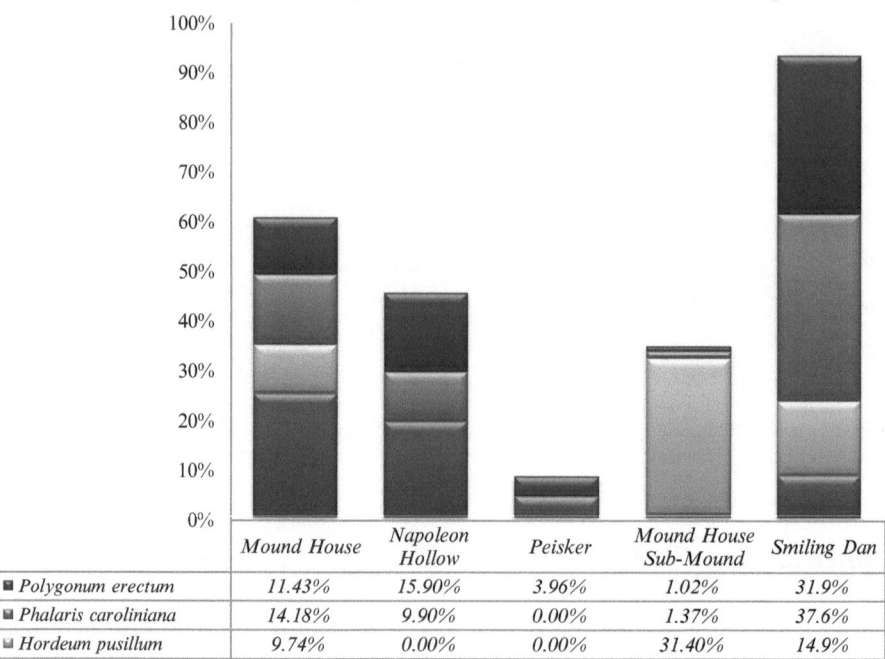

Eastern Agricultural Crop Seeds as % of Total Seed Assemblage

	Mound House	Napoleon Hollow	Peisker	Mound House Sub-Mound	Smiling Dan
■ *Polygonum erectum*	11.43%	15.90%	3.96%	1.02%	31.9%
■ *Phalaris caroliniana*	14.18%	9.90%	0.00%	1.37%	37.6%
■ *Hordeum pusillum*	9.74%	0.00%	0.00%	31.40%	14.9%
■ *Chenopodium berlandieri*	24.97%	19.40%	4.33%	0.68%	8.6%

Fig. 25 EAC crop seeds as proportion of the total seed assemblage (all identified seeds plus unidentifiables, but not seed fragments). The sub-mound assemblages from Peisker and Mound House, the midden assemblage from Napoleon Hollow, and the Feature 379 assemblage from Mound House represent different contexts associated with mound complexes; Smiling Dan is included for comparison (Asch and Asch 1985a, 1986; Staab 1984; Schroeder 1998). *Chenopod from Peisker identified only as "*Chenopodium* sp"

structures; the pits were shallow and full of redeposited midden; as at Mound House, none of these were considered storage pits by the excavators. Another curious similarity lies in the abundance of burned limestone at both sites. Limestone was used to temper some Middle Woodland ceramics, but Staab proposes that the abundance of limestone at Peisker indicates quick lime production. This is an intriguing possibility given the uses of lime for both construction and food processing and its possible symbolic uses as white paint or plaster (Staab 1984). There is also evidence of ceramic manufacture at Peisker, including caches of unfired clay and prepared hearths with unfired clay cups in them (Staab 1984: 90–97).

Crop seeds make up less than 10% of the seed assemblage from Peisker, an anomalously low proportion in comparison to every other Middle Woodland site in the valley (Fig. 25). The botanical and faunal remains from Peisker suggested to Staab that the "food quest" *was* a major activity for visitors/residents. Her conclusion may have been influenced by her conviction that agriculture was not practiced by Middle Woodland people and her seeming lack of familiarity with scholarship on the EAC (Staab 1984: 90), which caused her to deemphasize the significance of the almost complete lack of crop seeds at Peisker. Figure 25 illustrates this disparity in comparison to the sites previously discussed.

The lack of EAC seeds suggests that Peisker was even less likely to have been a site of agricultural activity than Mound House, but there are complicating factors. Almost half of the assemblage (418 of 808 seeds) was carbonized prostrate knot-weed (*Polygonum aviculare*), a species closely related to the crop seed erect knot-weed, whose native status has been considered enigmatic. The most recent reevaluation of the species (Costea and Tardif 2004) identified a native North American subspecies, *P. aviculare* ssp. *buxiforme*. Introgression between subspecies of this widespread plant is common, so the native subspecies is considered part of the *P. aviculare* complex, the remainder of which consists of weeds introduced by Euro-American settlers (Costea and Tardif 2004). Most government websites still list *P. aviculare* as an introduced species, but native *P. aviculare* ssp. *buxiforme* is still widespread in North America according to the United States Department of Agriculture.

The recently confirmed native status of *P. aviculare* necessitates a reevaluation of this plant's possible economic use among ancient North American people. Because of its aggressive growth habits and preference for packed earth, the distribution of *P.aviculare* globally is closely tied to human populations; it was likely present at Middle Woodland habitation sites (Costea and Tardif 2004). Middle Woodland people almost certainly would have recognized its similarity to erect knotweed, an important source of food (see Fig. 26). If so, it is possible that *P. aviculare* was also harvested for food, although it does not seem to be nearly as widely represented archaeologically. Alternatively, Middle Woodland people may have used *P. aviculare* for medicinal purposes, as did many historic North American peoples. This plant was used as a treatment for pain, inflammation, and diarrhea from the Pacific Northwest to New York and the southern Appalachians, and its medicinal uses continue to be explored by pharmacologists (Costea and Tardif 2004; Moerman 2009). The lone concentration of *P. aviculare* from the sub-mound features at Peisker is reminiscent of a smaller mass of little barley ($n=92$) from one of the sub-mound pits at Mound House: a single concentration of seeds in an otherwise nearly seed-free area. Given their location in otherwise seed-sterile sub-mound ritual areas, these two concentrations might be interpreted as burnt offerings. Since there is no evidence from elsewhere in the LIV that *P. aviculare* was a crop plant, it is not included in calculations of crop seed density, but its close relationship to erect knotweed makes this determination uncertain.

The second problem is that of comparability. The context of the Peisker assemblage is more similar to that of the sub-mound assemblage from Mound House (Schroeder 1998), than to the assemblage from Feature 379. There is a paucity of seeds in sub-mound contexts at both sites; the density of seeds from the sub-mound features at Mound House was only 0.7 seeds per liter, at Peisker only 0.03 seeds per liter, while in the Feature 379 samples from Mound House have a seed density of 6.4 seeds per liter. Thus, the overall proportion of crop seeds needs to be understood

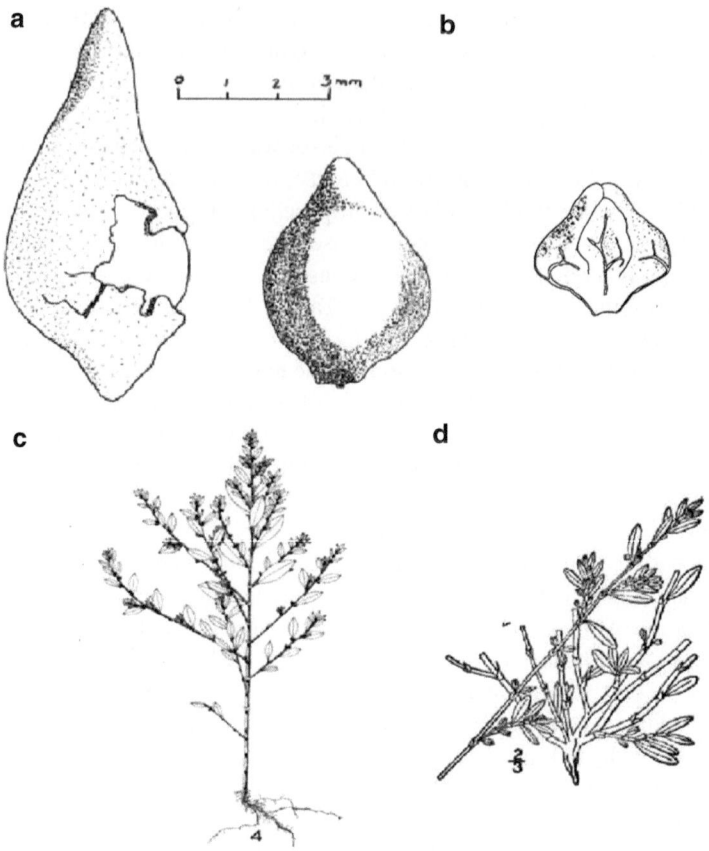

Fig. 26   (a) *Polygonum erectum* dimorphic achenes, from Washington University Paleoethnobotany Guide. (b) *Polygonum aviculare* ssp. *buxiforme* achene, from Costea and Tardif (2004: 486). (c) *Polygonum erectum* plant, from Steyermark (1981) *P. aviculare* ssp. *buxiforme* plant from Britton and Brown 1913

within the context of much sparser plant residues in general. With the exception of a single concentration of little barley at Mound House and *P. aviculare* at Peisker, no more than 5 of any one seed were recovered from the entire sub-mound excavations at either site. As Fig. 25 shows, the off-mound assemblage from Mound House Feature 379 has a lesser proportional frequency of crop seeds than the non-mound sites, but more than the sub-mound features. This suggests at least two functionally distinct areas at floodplain mound centers, and off-mound sampling at Peisker could greatly strengthen this conclusion. Had the midden deposits on the nearby sand bank at Peisker been sampled, they may have shown a seed density and distribution more similar to the Feature 379 assemblage.

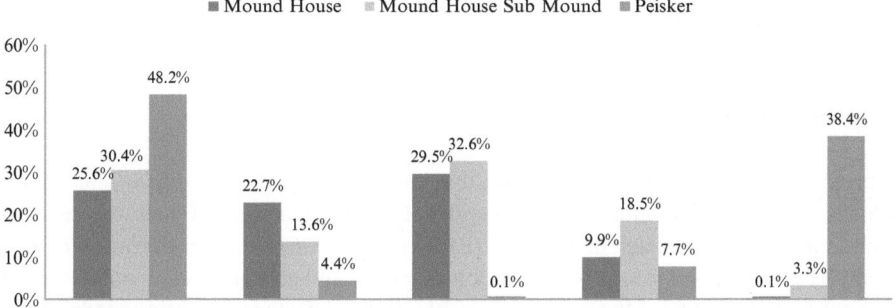

Fig. 27 Nutshell taxa as a proportion of the total nutshell assemblage, all fragments >2 mm (Schroeder 1998; Staab 1984)

The most obvious difference between the nutshell assemblages at Mound House and Peisker is the relative abundance of acorn at Peisker (Fig. 27). Staab suggests that if lime was produced at Peisker, it may have been useful for leaching tannins from acorns, although, historically, American Indians used lye or pure water for this purpose (Staab 1984: 98; Zawacki and Hausfater 1969: 39). This connection is thus tenuous. Evidently acorn processing took place at Peisker; acorn shells are better represented here than at any other site under analysis.

Faunal remains at Peisker were abundant and diverse, so Staab's assertion that the "food quest" was central to activity at Peisker may be accurate, if imprecise. Faunal analysis from Mound House is not yet available, but animal and fish bones are dense in all the heavy fractions I examined, and the proximity of the site to a backwater lake makes it likely that hunting and fishing were central to activity at the site. However, based on the available botanical assemblage, it does not seem that agricultural production was a part of the "food quest" at either site. As at Mound House, nuts may have been processed at the site.

Staab concluded that Peisker was inhabited probably in the fall, and possibly in the spring based on several lines of evidence. The most convincing is somewhat novel. A number of wasp nests with thatch impressions on them were recovered in the vicinity of frequently repositioned post-molds. This provides a line of evidence that the sub-mound structures were occupied only seasonally. Since people tend not to tolerate wasps in the walls of occupied buildings, the structures probably stood empty during at least part of the summer (Staab 1984: 109). The lack of summer season migratory bird bones provides another argument against a summer occupation (Staab 1984: 74).

Several lines of evidence suggest that ritual at Mound House was focused on flooding and world renewal and that visitation may have occurred in the late spring or early summer. It is at least possible, then, that Mound House and Peikser do not represent sites of formal visitation for separate communities, but rather different locations on a seasonal round with overlapping constituencies. As at Mound House, crop seeds are much less abundant at Peisker than at other habitations, but sampling from off-mound contexts would allow for a clearer picture of plant use at the site.

Napoleon Hollow

Current models of site function posit that most activity at bluff-top mound complexes was focused on mortuary ritual and that participants were of the same residential community or cluster of hamlets (Ruby et al. 2006). For the purposes of this analysis, though, it is just as likely that bluff-top mound complexes served as sites of formal visitation where the exchange of agricultural material and knowledge took place, but perhaps on a more local scale. The overall density of seeds at Napoleon Hollow was very low, only 0.09 seeds per liter—comparable to the seed density in the sub-mound features of Peisker (0.03 seeds per liter). But the context of the Napoleon Hollow assemblage is more similar to Feature 379 at Mound House because it was sampled from off-mound, extensive midden areas, not sub-mound features. Napoleon Hollow middens are more crop seed poor than Feature 379 and much more so than the other habitation sites in the valley (Fig. 25). Little barley is entirely absent.

Asch and Asch (1986) evaluate the significance of the difference in patterning of crop seeds between mound (Napoleon Hollow) and non-mound (Smiling Dan) sites. In so doing, they provide an alternative point of view on the problem which has been central to this analysis. They rule out problems with sampling and preservation as causes and posit three additional explanations: (1) the difference reflects different resource availability in the vicinity of the two sites; (2) multiple subsistence strategies were practiced in the valley, and inhabitants of Napoleon Hollow happened to rely more on game and fish, while those at Smiling Dan practiced agriculture; or (3) Napoleon Hollow was not a primary residence, and the range of activities that took place there was limited (Asch and Asch 1986: 500). They conclude that the third explanation is the most reasonable given their knowledge of LIV ecology and settlement and further speculate: "The paucity of nuts and cultivated seeds implies either that stores of them were depleted at the season of the site's occupation or that it was not deemed worthwhile to transport them from the central base camp" (1986: 506). This conclusion is logically flawed. If seeds were not harvested, processed, or stored at Napoleon Hollow, then their presence indicates the exact opposite: that they were purposefully brought to the site for consumption.

The case of Napoleon Hollow is somewhat different than that of Peisker and Mound House, because Napoleon Hollow is located in the same physiographic location as most Middle Woodland hamlets—the bluff base. For the Aschs' analysis, it was both correct and straightforward to eliminate ecology and resource availability as possible explanations for the variation in crop seed abundance they observed between Napoleon Hollow and Smiling Dan: the two site locations are nearly identical in terms of ecology and topography. For the floodplain mound centers, ecology and human behavior are not so easily disentangled. In fact, the original rationale for my supposition that Mound House was not a site of agricultural production was ecological. The frequently flooded and poorly drained soils surrounding Mound House would have been a risky and ill-suited environment for the type of agriculture practiced (presumably without flood control), considering what we know about the optimal conditions for the crops in question. But how to explain the presence of crop seeds in middens at such an unlikely location for agricultural production? I will now turn to a discussion of seed exchange networks, and the role Mound House may have played in maintaining seed security during the Middle Woodland period.

Conclusions: Mounds and Seed Exchange

The number of comparable assemblages in the relatively small area of the LIV gives us a window not only into what people ate but into how they used the landscape to produce it. The landscape that encompasses the sites described here is at a human scale: a person could easily travel from Massey to Peisker, the two sites most distant from each other, in 3 days of easy walking. The people who lived at these sites almost certainly visited each other and exchanged goods and knowledge. I argue that the exchange of both material and expertise would have been necessary to build and maintain a system of seed crop production.

If the most recent redating of LIV sites is accurate, then it supports the hypothesis that mound-building populations were slowly moving south from the central valley at the beginning of the Middle Woodland Period (King et al. 2011). However, the presence of artifacts with stylistic affiliations to the south suggests that other populations may have been moving into the valley or its tributaries at the same time (Farnsworth and Koski 1985: 131–159). To complicate matters further, there may have been some continuity with the much smaller Early Woodland population of the LIV. At Peisker, Early Woodland pottery is separated from Middle Woodland deposits by flood-borne sediments, but the earlier occupation still indicates some continuity in the use of the landscape between the Early and Middle Woodland (i.e., Peisker was visited by both Early and Middle Woodland people) and so perhaps continuity in population and cultural memory as well. However many populations coexisted in the valley; they simultaneously began producing seed crops and intensively harvesting hazelnuts at a scale that is archaeologically visible during the Middle Woodland period.

We know very little about the agricultural system Middle Woodland people employed, but to harvest seed crops, they would have had to minimally *clear land* and *save seeds*. Several analysts have addressed the issue of land clearance, using the prevalence of hazelnut and the composition of pollen cores to argue for burning (Asch and Asch 1985a; Delcourt and Delcourt 2004; Johannessen 1988; Simon and Parker 2006). The means by which farmers obtained seeds and knowledge about how to plant, harvest, and store them, on the other hand, has not been investigated. Erect knotweed and goosefoot had both been consumed in smaller amounts in and around the LIV for thousands of years, so the particulars of their preferred habitats, as well as harvesting times and methods, were probably common knowledge.

There is evidence here and elsewhere in the Eastern Woodlands for cultivation of seed crops hundreds of years earlier, including occasional finds of domesticated varieties of squash and sumpweed at Archaic sites. However, the botanical assemblages reported here are the earliest in the LIV to exhibit hundreds or thousands of crop seeds, so the intensity with which these resources were harvested increased greatly at this time. Little barley and maygrass were cultivated for the first time in the LIV during the Middle Woodland period. To grow these crops, farmers would have needed to know their preferred habitat and life cycle, but to facilitate larger harvests, information about how and where to store seeds and where and when to plant them would have been necessary.

The topics of seed security and seed selection are of considerable interest to anthropologists, crop geneticists, and development workers today. Seed security refers to the ability of individual farms to meet their needs for planting from year to year. For most crops, seeds have to be processed and stored differently for planting than they are for grain. Seeds stored for planting are sometimes destroyed by fires or pests over the winter. If farmers plant all of their saved seeds and then experience an early season drought, flood, or pest attack, they often do not have enough stored seed left for a second planting, causing them to seek inputs of seed from other farmers. Crop degeneration due to inbreeding in seeds saved on a single small farm can also lead farmers to exchange seed (Stromberg et al. 2010: 540).

Researchers involved in development initiatives are interested in knowing what networks farmers rely on to get more seed, so that they can make use of these networks to deliver aid. These studies focus on small-scale subsistence farmers with little market access, so they can provide useful insights into the needs and decisions of prehistoric farmers attempting to maintain a stable agricultural system. McGuire's (2008) study of Ethiopian sorghum farmers provides one example. In his study, there were a variety of factors that affected the likelihood that a farmer would need to obtain seed, including the location of his farm, the size of his holdings, and his access to draft animals. But no matter the reason for the shortage, farmers selected sources of seed based on *reliability*. When farmers were in need of seed, they tended to go to whatever location had a constant supply that could be handed over quickly, whether that was a neighbor, a family member, or a marketplace. By way of summary, McGuire (2008: 13) writes "membership in social networks matters a great deal for seed dissemination. A number of farmers referred to the ease or difficulty of social transactions with other farmers as a reason for preferring neighbors or markets, respectively."

A seed shortage is the kind of stochastic event that might have permanently disrupted the development of agricultural systems among dispersed prehistoric populations by forcing farmers to invest much of their time and energy in maintaining alternatives to crop production in case they ran out of seed stock. Arguably, it is only when obstacles to crop stability, such as seed shortages, are overcome that agricultural systems can emerge. In the Middle Woodland period, the ease or difficulty with which seed could be procured might mean the difference between planting crops and relying on wild resources, and over time the aggregate of such choices resulted in the emergence of an agricultural system. A place of formal visitation like Mound House, where people could count on coming together and exchanging material at a certain time of year, provides one answer to problems of farmer seed insecurity by providing a reliable place to exchange seed, especially for a dispersed and mobile population.

Studies in Oaxaca and Ethiopia have also focused on the dispersal of new varieties or novel crops among farmers (Badstue et al. 2006). Badstue et al.'s study of maize farmers in Oaxaca showed that seed exchanges there were less frequent than in the Ethiopian case, but that when they did occur trust was the most important determinant of farmers' choice of seed source. This was the case because farmers valued not only the seed but the information that was passed along with it: "Conversations with family members, compadres, and neighbors, as well as observations of what other farmers were growing, were among the most frequently reported ways of obtaining information about seed used elsewhere in the community" (Badstue et al. 2006: 265). Based on this evidence, a place where multiple communities came to live and prepare food together on a regular basis might be considered not just beneficial but *necessary* for the exchange of seed throughout the LIV, because it would have given farmers adequate time to develop relationships with, and examine the produce of, other farmers.

These cases are not presented to argue that Mound House was some kind of seed marketplace. Seed exchange might have been an incidental and intermittent part of activity at the site and yet had important beneficial results for the agricultural system of the valley. In many ethnographic cases, seed systems are informal and *ad hoc* because farmers only need to renew seed stocks occasionally. This results in seed systems that are built onto "pre-existing social networks that are not directly related to seed exchanges, such as community labor-sharing institutions" (Stromberg et al. 2010: 541), like those which must have existed for mound building during Middle Woodland times. Agriculture and mound building during the Middle Woodland were thus mutually reinforcing, with stored foods playing a role in sustaining communal labor and communal labor helping to maintain seed security. The idea that agricultural surplus facilitates the construction of monumental architecture is an old one. It is possible that monumental architecture can also support agricultural production, if it creates an impetus for routinized exchange.

Seed exchange might also have provided another avenue for social display or aggrandizement. McGuire (2008: 9) reports that socially prominent farmers are often known for giving seed to their neighbors and that even members of parliament sometimes serve as conduits of surplus seeds to their constituents. There is doubtless infinite variability in the norms governing seed saving and exchange worldwide and over time, yet other examples point toward a special role for individuals responsible for conserving and reproducing important seed varieties and disseminating them to the rest of the community. In the Democratic Republic of Congo, a recent study explored how "seed families" within agricultural communities served as custodians of particularly productive rice landraces. During times of stress, following droughts and wars, for example, seed families disseminated seed to a stable network of 5–10 households in return for labor or goods to be repaid later (Misiko 2010: 2).

The possibility that seeds may have been considered prestige goods on par with Hopewell exchange items has intriguing implications with regard to the gender dynamics of Middle Woodland societies. The following discussion of the possible gender dynamics of seed exchange rests on a fundamental assumption shared by most researchers of prehistoric eastern North America: women were primarily responsible for farming. This assumption is grounded in the universal testimony of the historic and ethnographic record to the role of women as farmers in North American societies, and especially as seed savers and planters (Watson and Kennedy 1991; Fritz 1999). During Middle Woodland times in the LIV, mortuary evidence indicates that women wielded less social power than men. Women were almost never buried in floodplain mounds and were less often interred with items of Hopewell exchange (Buikstra 1976). Charles (1995: 89) contends that "Competition among elites of different communities, and among lineages vying for status within communities, was played out in mortuary rituals and through the so-called Hopewell Interaction Sphere ... The large floodplain centers ... may have been the ultimate outcome of such competition." If competition for elite status was primarily con-tested via the exchange and internment of exotic Hopewell goods, then (based on their graves) women played relatively little role in Middle Woodland ritual politics. If exchange of Eastern Agricultural Complex seeds was also a venue for aggran-dizement and an instrument of social integration, as the ethnographic record of seed exchange suggests, then women may have played a greater role in Middle Woodland ritual and politics than previously supposed.

Konigsberg and Buikstra's (1995) analysis of male and female skeletal covari-ance at several LIV mound sites showed that women from the same burial commu-nity had a higher level of cranial variation than men, indicating greater residential mobility on the part of women. They suggest that the Middle Woodland societies of the LIV were patrilocal (1995:200–201). If so, the kinship networks of women would have been more geographically dispersed than those of their partners, fathers, and brothers. Hart (2001) has suggested that the evolution of matrilocal societies during the Late Woodland period facilitated the development of hardy maize varieties because generations of women living in the same location were able to develop seed stocks that were exceptionally well adapted to their agroecological situation. If women were responsible for saving and planting seeds in ancient societies as they were in historical ones, there would certainly have been advantages to matrilocality with regard to agriculture because women would have benefited from a lifelong apprenticeship with their mothers and grandmothers and seed stocks would have been continuously grown in the agroecological setting for which they were selected. But Middle Woodland societies were most likely patrilocal, making locations like Mound House where several residential communities (re)united crucial to the main-tenance of communication and exchange between dispersed female relatives.

Without denying the benefits of matrilocality to female farmers, it is clear that women with extensive social networks would have enjoyed a different suite of advantages when it came to maintaining productive seed stocks. The availability and genetic diversity of seed is a critical factor to any agricultural system. Hart (1999) discusses the implications of very small introductory populations of maize on the trajectory of that crop's spread throughout the Eastern Woodlands. The case for maize is somewhat different than for the EAC crops because the wild progenitors and/or wild populations of the EAC crops were widely distributed throughout the Eastern Woodlands (although maygrass did not grow wild within hundreds of miles of the LIV). However, the evolutionary issues he discusses with regard to the survival of maize as a population can also provide a useful framework for understanding the development of the Eastern Agricultural Complex. As Hart points out, introductory events need to coincide with subsequent gene flow. If founder populations are isolated, inbreeding and extinction can occur (1999: 153). This argument is applicable to the introduction of maygrass to areas like the LIV during the Middle Woodland—seed exchange networks like the one envisioned here would have been crucial to that introduction.

For human populations on the move, as some of those who settled in the LIV at the beginning of the Middle Woodland Period almost certainly were, Hart's discussion of fitness is also pertinent. He observes that crop varieties are optimized for the particular agroecology in which they are bred. When successful cultivars are planted in a different ecological setting than that in which they were developed, they are initially less fit and so less productive (Hart 1999: 153–154). The ubiquity of the Eastern Agricultural Complex crops across the midwest and south during the Middle Woodland attests to the fact that the problems of population isolation and variable agroecologies were overcome by the interconnectedness of populations—populations of people and through them populations of crop plants. Farmers were able to acquire successful cultivars as quickly as they encountered new agroecologies and were able to maintain these cultivars despite the challenges of inbreeding and seed loss.

The Eastern Agriculture Complex is extremely variable in its expression throughout the Eastern Woodlands and, like the seed systems that support subsistence agriculture today, the networks by which it was spread were no doubt diverse. Certainly mound centers are not necessary to the emergence of agricultural systems, and not all Middle Woodland mound centers are necessarily implicated. This analysis has attempted to make an argument based on the *relative scarcity* of seed crops and their spatial patterning at mound sites that the mound centers of the LIV were not sites of agricultural production. Paradoxically, the lack of seeds at the mound centers may be a clue to the important role they served in the agricultural system. If seed crops were not harvested, processed, or stored at these sites, their presence there indicates that they were brought to the mounds specifically for consumption or exchange. This argument is based on both ecological and archaeological evidence from three mound and three non-mound habitations in the LIV.

Given the importance of seed exchange, not only for the stability of agricultural systems, but also for the improvement of crop varieties, a logical next step would be an examination of seed morphology and variability. Seeds at sites such as Smiling Dan, where harvests were processed and stored, presumably represent a random sample of all seeds harvested. In contrast, if stored seeds were transported to sites like Mound House, they may have been deliberately selected for superior quality, either for feasting or exchange. The preliminary analysis of goosefoot and erect knotweed morphology in this analysis does not contradict the argument presented here, but neither does it provide a clear indication of unique seed morphology at Mound House in comparison to sites where comparable measurements have been taken.

A larger and more diverse botanical assemblage from both Mound House and Peisker might also strengthen my hypotheses, or refute them. The Mound House assemblage is still much smaller than all of the others analyzed. I used a resampling method based on rarefaction analysis to test the statistical significance of crop seed abundance between Mound House and Smiling Dan and showed that the relative scarcity of crop seeds at Mound House was probably not a result of sampling error. Still, other differences between the assemblages, for instance, the apparent abundance of summer fruit seeds with ritual and medicinal uses at Mound House, were not statistically significant. A larger sample has the potential to bring the distinctions between the assemblages into clearer focus. At Peisker, all excavations targeted mounds or sub-mound features, so that the areas where visitors would have lived, cooked, and informally exchanged knowledge and material remain unexamined.

The presence of archaeological correlates associated with an emergent agricultural system, such as abundant crop seeds, possible fallow field crops, storage facilities, processing tools, and improved cook wares (Smith 1992a: 207–209), all testify to the success of Middle Woodland farmers in overcoming the problem of seed insecurity. Without mechanisms for the dissemination of seeds and knowledge, agriculture is risky and unpredictable. Even for today's subsistence farmers, who exist in a much more ecologically constrained world, seed insecurity and breakdowns in networks of mutual support can lead farmers to abandon the cultivation of crops altogether (McGuire 2008). Based on this analysis, I suggest that mound complexes may have played a role in nurturing the emergence of an archaeologically visible agricultural system during the Middle Woodland period.

References

Anderson E (1952) Plants, man and life. Missouri Botanical Garden, St. Louis

Asch DL (1976) The Middle Woodland population of the lower Illinois valley: a study in paleodemographic methods. Northwestern University archaeological program scientific papers 1 Evanstan, IL

Asch DL, Asch NB (1985a) Archaeobotany. In: Stafford BD, Sant MB (eds) Smiling dan: structure and function at a Middle Woodlands settlement in the lower Illinois river valley. Center for American Archaeology, Kampsville, pp 327–400

Asch DL, Asch NB (1985b) Archaeobotany. In: Conner MD (ed) The Hill Creek Homestead and the Late Mississippian settlement in the lower Illinois river valley, vol 1. Kampsville Archaeological Center, The Center for American Archaeology, Kampsville, pp 115–170

Asch DL, Asch NB (1985c) Archaeobotany. In: Farnsworth KB, Walthall JA (eds) Massey and archie: a study of two hopewellian homesteads in the western Illinois uplands, Research Series, vol 3, Center for American Archaeology. Kampsville, IL, pp 327–401

Asch DL, Asch NB (1985d) Prehistoric plant cultivation in west central Illinois. In: Ford RI (ed) Prehistoric food production in North America, vol 75, Museum of anthropology. University of Michigan, Ann Arbor, pp 149–203

Asch DL, Farnsworth KB, Asch NB (1979) Woodland subsistence and settlement in west central Illinois. In: Brose DS, Greber N'o (eds) Hopewell Archeaology: the Chillicothe Conference. Kent State University Press, Kent, pp 80–85

Asch NB, Asch DL (1986) Woodland period archeobotany of the Napoleon hollow site. In: Wiant MD, McGimsey CR (eds) Woodland period occupations of the napoleon hollow site in the lower Illinois valley. Center for American Archeology, Kampsville, IL, pp 427–512

Badstue LB, Bellon MR, Berthaud J, Juàrez X, Rosas IM, Solano AM, Ramírez A (2006) Examining the role of collective action in an informal seed system: a case study from the central valleys of Oaxaca, Mexico. Hum Ecol 34(2):249–273

Blackman BK, Rasmussen DA, Strasburg JL, Raduski AR, Burke JM, Knapp SJ, Michaels SD, Rieseberg LH (2011) Contributions of flowering time genes to sunflower domestication and improvement. Genetics 187:271–287

Bohrer VL (1991) Recently recognized cultivated and encouraged plants among the hohokam. Kiva 56:227–235

Borojevic K (1994) Rhus spp.: Sumacs. electronic document, http://artsci.wustl.edu/~gjfritz/Rhus_spp.html. Laboratory Guide to Archaeological Plant Remains from Eastern North America. Accessed 1 Feb 2012

Braun DP (1979) Illinois hopewell burial practices and social organization: a reexamination of the Klunk-Gibson Mound Group. In: Brose DS, Greber N'o (eds) Hopewell archaeology. Kent State University Press, Kent, pp 66–79

Britton NL, Brown A (1913) An illustrated Flora of the northern United States, Canada and the British possessions: from Newfoundland to the parallel of the Southern boundary of Virginia, and from the Atlantic Ocean Westward to the 102d meridian, 2nd edn. Charles Scribner's, New York

Brose DS (1979) A speculative model of the role of exchange in the prehistory of the eastern Woodlands. In: Brose DS, Greber N'o (eds) Hopewell archaeology: the chillicothe conference. Kent State University Press, Kent, pp 3–8

Buikstra JE (1976) Hopewell in the lower Illinois valley: a regional approach to the study of human biological variability and prehistoric behavior. Northwestern University archaeological program scientific papers 2, Evanstan, IL

Buikstra JE (1979) Contributions of physical anthropologists to the concept of hopewell: a historical perspective. In: Brose DS, Greber N'o (eds) Hopewell archaeology. Kent State University Press, Kent, pp 211–219

Buikstra JE, Charles DK, Rakita GFM (1998) Staging ritual: hopewell ceremonialism at the Mound House site, Greene County, Illinois. Kampsville Studies in Archeology and History 1. Center for American Archaeology, Kampsville, IL

Burke JM, Tang S, Knapp SJ, Rieseberg LH (2002) Genetic analysis of sunflower domestication. Genetics 161:1257–1267

Caldwell JR (1964) Interacting spheres in prehistory. In: Caldwell JR, Hall RL (ed) Hopewellian studies: the chillicothe conference, Illinois State museum paper no. 12. Illinois State Museum, Springfield, IL. pp. 133–143

Calentine L (2005) The spoon toe site (11MG179): Middle Woodland gardening in the lower Illinois river valley. Unpublished PhD dissertation. Department of Anthropology, University of Missouri, Columbia

Carr C (2006a) Historical insights into the directions and limitations of recent research on Hopewell. In: Carr C, Tory Case D (eds) Gathering hopewell: society, ritual and ritual interaction. Springer, New York, pp 51–73

Carr C (2006b) Rethinking interregional hopewellian "Interaction". In: Carr C, Troy Case D (eds) Gathering hopewell: society, ritual, and ritual interaction. Springer, New York, pp 575–623

Charles DK (1995) Diachronic regional social dynamics: mortuary sites in the Illinois valley/ American bottom region. In: Beck LA (ed) Regional approaches to mortuary analysis. Plenum Press, New York, pp 77–100

Charles DK, Buikstra JE (eds) (2006) Recreating hopewell. University Press of Florida, Gainesville

Charles DK, Van Nest J, Buikstra JE (2004) From the Earth: minerals and meaning in the hopewellian world. In: Boivin N, Owoc MA (eds) Soils, stones and symbols: cultural perceptions of the mineral world. UCL Press, London, pp 43–70

Costea M, Tardif FJ (2004) The biology of canadian weeds: *Polygonum aviculare* L. Can J Plant Sci 131:481–506

Cowan WC (1985) Understanding the evolution of plant husbandry in eastern north America: lessons from the botany, ethnography and archaeology. In: Ford RI (ed) Prehistoric food production in north America, vol. 75. Museum of anthropology, University of Michigan, Ann Arbor, MI

Wet D, Jan M, Harlan JR (1975) Weeds and domesticates: evolution in a man-made habitat. Econ Bot 29(2):99–108

Delcourt PA, Delcourt HR (2004) Prehistoric native Americans and ecological change: human ecosystems in eastern north America since the pleistocene. Cambridge University Press, Cambridge

Dragoo DW (1964) The development of the adena culture and its role in the formation of Ohio hopewell. In: Caldwell JR, Hall RL (ed) Hopewellian studies, Scientific Papers, vol. 7. Illinois State Museum, Springfield, IL

Farnsworth KB (1990) The evidence for specialized middle woodland camps in western Illinois. Illinois Archaeologist 2(1&2):109–132

Farnsworth KB, Asch DL (1986) Early woodland chronology, artifact styles, and settlement distribution in the lower Illinois valley region. In: Farnsworth KB, Emerson TE. Early woodland archaeology, Kampsville seminars in archaeology, vol. 2. Center for American Archaeology, Kampsville, IL

Farnsworth KB, Koski AL (1985) Ceramics. In: Farnsworth KB, Koski AL (ed) Massey and Archie: a study of two hopewellian homesteads in the western Illinois uplands. Research Series, vol. 3. Center of American Archeology, Kampsville, IL. pp. 124–161

Fie SM (2006) Visiting in the interaction sphere: ceramic exchange and interaction in the lower Illinois valley. In: Charles DK, Buikstra JE (eds) Recreating hopewell. University of Florida Press, Gainesville, pp 427–445

Ford RI (1979) Gathering and gardening: trends and consequences of hopewell subsistence strategies. In: Brose DS, Greber N'o (eds) Hopewell archaeology: the chillicothe conference. Kent State University Press, Kent, pp 234–238

Fritz GJ (1990) Multiple pathways to farming in eastern precontact North America. J World Prehist 4(4):387–436

Fritz GJ (1999) Gender and the early cultivation of gourds in eastern North America. Am Antiq 64(3):417–429

Fritz GJ (2010) The meaning of maygrass. Paper presented at the 34th Annual Meeting of the Society of Ethnobiology, Columbus, OH

Fritz GJ, Smith BD (1988) Old collection and new technology: documenting the domestication of chenopodium in eastern North America. MidCont J Archaeol 13(1):3–27

Gasser RE (1982) Hohokam use of desert plant foods. Desert Plants 3:216–234

Gilmore MR (1977) Uses of plants by Indians of the Missouri river region. University of Nebraska Press, Lincoln

Gremillion KJ (1993) Crop and weed production in eastern North America: the chenopodium example. Am Antiq 58(3):496–509

Griffin JB (1952) Culture periods in the eastern United States prehistory. In: Griffin JB (ed) Archaeology of the Eastern United States. University of Chicago Press, Chicago

Hall RL (1979) In search of the ideology of the adena-hopewell climax. In: Brose DS, Greber N (eds) Hopewell archaeology: the chillicothe conference. Kent State University Press, Kent, pp 258–265

Harlan JR, de Wet JM (1965) Some thoughts on weeds. Econ Bot 18(1):16–24

Hart JP (1999) Maize agriculture evolution in the eastern woodlands of North America: a Darwinian perspective. J Archaeol Method Theory 6(2):137–180

Hart JP (2001) Maize, matrilocality, migration and northern Iroquoian evolution. J Archaeol Method Theory 8(2):151–182

Johannessen S (1988) Plant remains and culture change: are paleoethnobotanical data better than we think? In: Hastorf CA, Popper VS (eds) Current paleoethnobotany: analytical methods and cultural interpretations of archaeological plant remains. University of Chicago Press, Chicago, pp 145–166

King JL, Buikstra JE, Charles DK (2011) Time and archaeological traditions in the lower Illinois valley. Am Antiq 76(3):500–528

King JL, Rudolf KZ, Buikstra JE (2010) Habitation occupations at floodplain mound sites in the lower Illinois valley: evidence from the mound house site (11GE7). Paper presented at the 55th annual meeting of the midwest archaeological conference, Bloomington, IN

Konigsberg LW, Buikstra JE (1995) Regional approaches to the investigation of past human bio-cultural structure. In: Beck LA (ed) Regional approaches to mortuary analysis. Interdisciplinary Studies in Archaeology. Plenum Press, New York, pp 191–205

Lopinot NH, Fritz GJ, Kelly JE (1991) The archaeological context and significance of polygonum erectum achene masses from the American bottom region. Paper presented at 14th annual meeting of the Society of Ethnobiology, St. Louis, MO

Martin GJ (1995) Ethnobotany: a methods manual. Chapman and Hall, London

McGimsey CR, Wiant MD (1986) The woodland occupations: summary and conclusions. In: Wiant MD, McGimsey CR (eds) Woodland period occupations of the Napoleon hollow site in the lower Illinois valley. Center for American Archeology, Kampsville, IL

McGregor JC (1958) The pool and irving sites: a study of hopewell occupations in the Illinois river valley. University of Illinois Press, Urbana

McGuire SJ (2008) Securing access to seed: social relations and sorghum seed exchange in eastern Ethiopia. Hum Ecol 36(2):217–229

Misiko M (2010) Indigenous seed institutions in fragile communities. In Second African Rice Conference. Innovations and Partnerships to Realize Africa's Rice Potential, Bamako, Mali

Moerman DE (2009) Native American medicinal plants: an ethnobotanical dictionary. Timber Press, Portland

Montgomery FH (1977) Seeds and fruits of plants of eastern Canada and northeastern. University of Toronto Press, Toronto

Moorehead WK (1892) Primitive man in Ohio. G.P. Putnam, New York

Munson PJ (1984) Weedy plant communities on the mud-flats and other disturbed habitats in the central Illinois river valley. In: Munson PJ (ed) Experiments and observations on aboriginal wild plant utilization in eastern North America. Indiana Historical Society, Indianapolis

Murray PM, Sheehan MC (1984) Pehistoric polygonum use in the Midwestern United States. In: Munson PJ (ed) Experiments and observations on aboriginal wild plant food utilization in eastern North America. Indian Historical Society, Indianapolis

Potter LD, Ross Moir D (1961) Phytosociological study of burned deciduous woods, turtle mountains North Dakota. Ecology 42(3):468–480

Ruby BJ, Carr C, Charles DK (2006) Community organization in the Scioto, Mann, and Havana hopewellian regions: a comparative perspective. In: Carr C, Troy Case D (ed) Gathering hopewell: society, ritual, and ritual interaction. Springer, New York

Sauer JD (1950) Pokeweed, an old American herb. Ann Mo Bot Gard 38:83–88

Scarry CM (1993) Introduction. In: Margaret Scarry C (ed) Foraging and farming in the eastern woodlands. University of Florida Press, Gainesville, pp 3–12

Schroeder M (1998) Archaeobotanical analysis. In: Buikstra JE, Charles DK, Rakita GFM (ed) *Staging* Ritual: hopewell ceremonialism at the mound house site, Greene county, Illinois, Kampsville studies in archeology and history, vol. 1. Center for American Archeology, Kampsville pp. 180–188

Seeman MF, Wilson HD (1984) The food potential of chenopodium for the prehistoric midwest. In: Munson PJ (ed) Experiments and observations on Aboriginal wild plant food utilization in eastern North America, Prehistory research series, vol 6. Indiana Historical Society, Indianapolis, pp 299–316

Shetrone HC (1920) The culture problem in Ohio archaeology. Am Anthropol 22:144–171

Simberloff D (1972) Properties of the rarefaction diversity measurement. Am Nat 106(949): 414–418

Simon ML, Parker KE (2006) Prehistoric plant use in the American bottom: new thoughts and interpretations. Southeast Archaeol 25(2):212–257

Smith BD, Yarnell RA (2009) Initial formation of an indigenous crop complex in eastern North America at 3800 B.P. Proc Natl Acad Sci U S A 106(16):6561–6566

Smith BD (1992a) Rivers of change: essays on agriculture in eastern North America. Smithsonian Institution Press, Washington

Smith BD (1992b) The role of chenopodium as a domesticate in premaize garden systems of the Eastern United States. In: Smith BD (ed) Rivers of change: essays of early agriculture in eastern North America. Smithsonian Institution Press, Washington, pp 103–131

Smith BD, Funk VA (1985) A newly described subfossil cultivar of *chenopodium*: (Chenopodiaceae). Phytologia 57(7):445–448

Staab ML (1984) Peisker: an examination of middle woodland site function in the lower Illinois river valley. Ph.D. dissertation, Department of Anthropology, University of Iowa, Iowa City

Stafford BD, Sant MB (eds) (1985) Smiling Dan: structure and function at a middle woodland settlement in the Illinois valley, 2. Center for American archaeology, Kampsville, IL

Steyermark JA (1981) Flora of Missouri, 6th edn. The Iowa State University Press, Ames

Stromberg PM, Pascual U, Bellon MR (2010) Seed systems and farmers' seed choices: the case of maize in the Peruvian Amazon. Hum Ecol 38:539–553

Struever S (1964) The hopewell interaction sphere in riverine-western great lakes culture history. In: Caldwell JR, Hall RL (ed) Hopewellian studies, Illinois state museum papers no. 12. Illinois state museum, Springfield, IL, pp. 85–106

Struever S (1968) Woodland subsistence-settlement systems in the lower Illinois valley. In: Binford SR, Binford LR (eds) New perspectives in archaeology. Aldine, Chicago

Struever S, Houart GL (1972) An analysis of the hopewell interaction sphere. In: EN Wilmsen EN (ed) Social exchange and interaction. Anthropological papers of the museum of anthropology, vol. 46. University of Michigan, Ann Arbor. pp. 47–80

Turner LM (1934) Grassland in the floodplain of the Illinois rivers. Am Midl Nat 15(6):770–780

Van Nest J (2006) Some ethnogeological aspects of the Illinois valley hopewell mounds. In: Clark DK, Buikstra JE (eds) Recreating hopewell. University Press of Florida, Gainesville, pp 402–426

Watson PJ, Kennedy MC (1991) The development of horticulture in the eastern woodlands of North America. In: Gero J, Conkey MW (eds) Engendering archaeology. Basil Blackwell, Oxford, pp 255–275

Wiant MD, Farnsworth KB, Hajic ER (2009) The archaic period in the lower Illinois river basin. In: The archaic: diversity and complexity across the midcontinent. SUNY Press, Albany

Wills DM, Burke JM (2006) Chloroplast DNA variation confirms a single origin of domesticated sunflower (*Helianthus annuus* L.). J Hered 97(4):403–408

Wilson HD (1981) Domesticated *Chenopodium* of the ozark bluff dwellers. Econ Bot 35:233–239

Wilson HD, Heiser CB Jr (1981) The origin and evolutionary relationships of "Huauzontle" (*Chenopodium nuttalliae* stafford), domesticated chenopod of Mexico. Am J Bot 66:198–206

Wobst ER (1974) Boundary conditions for paleolithic social systems: a simulation approach. Am Antiq 39:147–178

Wray DE (1952) Archaeology of the Illinois valley:1950. In: Griffin JB (ed) Archaeology of the eastern United States. University of Chicago Press, Chicago, pp 152–164

Yarnell RA (1969) Contents of human paleofeces. In: Watson PJ (ed) The prehistory of Salts Cave, Kentucky. Reports of investigations, vol. 16. Illinois State Museum, Springfield, IL pp. 41–54

Yarnell RA (1972) *Iva annua var. macrocarpa*: extinct American cultigen. Am Anthropol 74:335–341

Zawacki AA, Hausfater G (1969) Early vegetation of the lower Illinois valley: a study of the distribution of floral resources with reference to prehistoric cultural-ecological application. Illinois valley archaeological program 1. Illinois State Museum, Springfield, IL

Appendix

N.G. Mueller, *Mound Centers and Seed Security: SpringerBriefs in Plant Science,*
DOI 10.1007/978-1-4614-5921-7, © Springer Science+Business Media New York 2013

Lab number	21	30	37	45	47	57	58	60	61	63	94	95	66	70	73	75	76	Total
Volume (L)	10	6	6	7	4.5	10	10	11.5	10	10	9	10	2	10	10	3.5	10	139.5
Nutshell weight																		
Carya sp. (Thick)	0.14	0.41	0.23	0.05	0.30	0.11	0.04	1.3	0.8	0.28	0.01	1.21	0.3	0.18		0.15	0.42	5.93
Carya sp. (Thin)	0	0	0	0.02			0.01		0.07	0.05		0						0.15
Juglans nigra	0	0	0	0		5.87		3.35	3.6	0.09	0.16	0		0.27				13.52
Juglandaceae	0.01	0.09	0	0	0.13	1.27		1.46	0.42	0.01	0.02	0.52	0.09	0.14		0.01		4.17
Corylus americana	0.06	0.11	0	0.01		0.58		0.29	0.11	0.01	0.02	0.42	0.07	0.05		0.04	0.07	1.84
Quercus sp.	0	0	0	0		0.01						0						0.01
Nutshell, indet.	0	0	0	0	0.01	0.51		0.06	0.64	0.01		0	0.09	0.05				1.37
Nutshell count																		
Carya sp. (Thick)	8	26	3	7	19	2	1	75	41	21	2	86	22	11		9	30	363
Carya sp. (Thin)	0	0	0	3			1		2	2		0						8
Juglans nigra	0	0	0	0		170		95	41	1	2	0	5	8				322
Juglandaceae	1	6	0	0	16	135		141	29	2	3	58	7	20		1		419
Corylus americana	4	7	0	1		34		24	11	1	3	32	5	7		4	8	141
Quercus sp.	0	0	0	0		1						0						1
Nutshell, indet.	0	0	0	0	1	66		6	67	4		0	10	10				164
Eastern agricultural complex																		
Chenopodium berlandieri				15		10	8	3	1		192	2	3	2				236
Hordeum pusillum				1	2	7	7	6	2	3		39		6		7	12	92

	1	2	3	4	5	6	7	8	9	10	11	12	13	14	15	16	Total
Phalaris caroliana		1	1	12	2	19	17	10	9	1	43	9	2	6	0	3	134
olygon um erect		1	2	5	1	21	17	14	9	2	33	0	0	0	0	2	108
Total EAC		1	3	33	5	57	49	33	21	6	268	50	5	14	7	17	570
Other seeds																	
Amaranthus sp.											1		1				2
Diospyros virginiana			2								7						11
Echinocloa sp.								1									1
Festucoid													1				1
Galium sp.		1						1						1			3
Nelumbo lutea										1	2						3
Panicoid			6			55	28	18	9	1	1			7			126
Phytolacca americana		1									1						3
Poaceae			2			8	2	2	1		2			1		6	11
Polygon um biconvex									3	1	1			1			17
Portulaca oleracea			6			10	6	6			12	15	1	2		2	59
Rhus sp.			15		1	3		1			16	11				1	53
Rubus sp.						1											1
Solanum sp.			2			2		1			6	10	3	1			25
Vaccinium sp.			1														1
Verbena sp.			1														1
Vitis spp.	1		2			6		15	3		15	3	4	3		4	36
Unidentified seed			2			2	3	2	1		15	6	4	3	1	1	39
Total seeds	2	2	12	63	7	131	94	73	43	12	314	92	15	27	7	30	924